Mobile Communication and Greater China

This edited volume is the first book-length study focusing entirely on mobile phone use in China. Drawing on examples from a wide range of contemporary situations in China and beyond, the contributors argue that the mobile phone is in fact an important means by which one can understand a rapidly changing China, and the developing culture of mobile phone usage reflects both the cultural norms and struggle of the people.

Through a theoretical comparison of usage in the West and in China, the editors assert the uniqueness of China's experience, highlighting that Chinese society is being exposed simultaneously to a rapid process of industrialization and cyberization. The contributors maintain that such density of experience under a compressed period combined with a thick cultural heritage and a country still under a dictating rule provides a unique situation and offers deep insights into Chinese culture in general.

This work will be of great interest to all students and scholars of Asian communication studies, ICT and Chinese culture and society.

Rodney Wai-chi Chu is currently Assistant Professor in the Department of Applied Social Sciences, The Hong Kong Polytechnic University.

Leopoldina Fortunati is Professor of Sociology of Communication and Sociology of Cultural Processes in the Faculty of Educational Sciences at the University of Udine.

Pui-lam Law is currently Assistant Professor in the Department of Applied Social Sciences, The Hong Kong Polytechnic University and visiting research fellow at the Centre for Creative Industries Studies of Peking University.

Shanhua Yang is Professor in the Department of Sociology at Peking University.

Routledge research on social work, social policy and social development in Greater China

Series editors: Angelina Woon-ki Yuen-Tsang, Ben Hok-bun Ku and Ngai Pun

Hong Kong Polytechnic University

This series focuses on issues related to the rapid social change in contemporary Mainland China, Hong Kong, Macao and Taiwan. Comprising contributions from diverse disciplines, the series will explore and respond to the impacts of marketization, globalization, urbanization, modernization, migration, technological change, consumerism, dualism and disasters on people's livelihoods, well-being and social life in Chinese societies.

1. Mobile Communication and Greater China

Edited by Rodney Wai-chi Chu, Leopoldina Fortunati, Pui-lam Law and Shanhua Yang

Mobile Communication and Greater China

Edited by
Rodney Wai-chi Chu, Leopoldina Fortunati, Pui-lam Law and Shanhua Yang

Routledge
Taylor & Francis Group
LONDON AND NEW YORK

First published 2012
by Routledge
2 Park Square, Milton Park, Abingdon, Oxon OX14 4RN

Simultaneously published in the USA and Canada
by Routledge
711 Third Avenue, New York, NY 10017

Routledge is an imprint of the Taylor & Francis Group, an informa business

British Library Cataloguing in Publication Data
A catalogue record for this book is available from the British Library

Library of Congress Cataloging in Publication Data
A catalog record for this book has been requested

ISBN 978-0-415-67871-1 (hbk)
ISBN 978-0-203-12094-1 (ebk)

Typeset in Times New Roman
by Book Now Ltd, London

MIX
Paper from responsible sources
FSC www.fsc.org FSC® C004839

Printed and bound in Great Britain by
CPI Antony Rowe, Chippenham, Wiltshire

Contents

Illustrations

Figures

Tables

Notes on contributors

Chung-tai Cheng gained his PhD in Sociology from the Peking University in China. He is currently Instructor in the Department of Applied Social Sciences, The Hong Kong Polytechnic University. His research interest is in the topic of the normative implications of ICT uses in China, especially in relationship to self-identification and social relationships. His presentation 'Imagined performativity: the great virtue of cyberspace in contemporary Chinese workers' social lives' won the Best Presentation Award in *COST Action 298 Conference: The Good, the Bad, and the Challenging* in the year 2009. He has published several journal articles and book chapters and presented conference papers, mainly focusing on the social and cultural impacts of ICTs.

Rodney Wai-chi Chu obtained his BA in Social and Political Science from the University of Cambridge, and his MPhil and PhD in Sociology from the Chinese University of Hong Kong and the University of Queensland, respectively. He is Assistant Professor in the Department of Applied Social Sciences, The Hong Kong Polytechnic University. Currently, he is focusing on research concerning the socio-cultural impact of information and communication technologies (ICTs) on contemporary Chinese, with particular reference to the use of mobile telephony and the Internet by migrant workers and university students. His recent publications concerning the mobile phone study include 'Mobile Phones and New Migrant Workers in a South China Village' [with Shanhua Yang] (*New Technologies in Global Societies*, World Scientific, 2006), 'The Dynamics of Cyber China: The Characteristics of Chinese ICT Use' (*Journal of Knowledge, Technology and Policy*, Vol. 21, No. 1, 2008) and 'Cultural Convulsions: Examining the Chineseness of Cyber China' [with Chung-tai Cheng] (*Online Society in China*, Routledge, 2011). He is also the co-guest editor of two special issues on the China mobile phone study for the *Journal of Knowledge, Technology and Policy*, Vol. 21, Nos. 2 & 3, 2008.

Silvia Elaluf-Calderwood gained her BSc from the Catholic University of Peru, her MSc in Computer Engineering from Imperial College London and her PhD in Information Systems from the London School of Economics and Political Science. She has worked in the telecommunications industry in the UK and the Netherlands. She has a strong technical and managerial background in the

Internet and Telecoms business. Her area of speciality is mobility, with a particular area of interest in the changes in working practices that mobile technology presents to mobile workers, challenging their definitions of identity and trust relationships at work. She has published a number of papers in her field of research and a book chapter in *Agile Information Systems* (Elsevier, 2006). She is engaged in working for the Mobility group at the LSE (http://mobility.lse.ac.uk) in the MVCE (Mobile Virtual Centre of Excellence) in the UK and in consultancy projects for the NOKIA corporation.

Leopoldina Fortunati is Professor of Sociology of Communication and Sociology of Cultural Processes in the Faculty of Educational Sciences at the University of Udine. She has conducted several research projects in the field of gender studies, cultural processes and communication technologies. She is the author of many books and was the editor of *Mediating the Human Body: Technology, Communication and Fashion* (Lawrence Erlbaum Associates, 2003) with J. Katz and R. Riccini and co-edited the volume *New Technologies in Global Societies* (2006). She has published extensively in important journals. She represents Italy in the COST Domain Committee 'Individuals, Societies, Cultures and Health' and in the COST 298 'Participation in the Broadband Society'. She was involved in the European research project SIGIS, 'Strategies of Inclusion: Gender and the Information Society', and COST248, 'The Future European Telecommunications User', and was vice-chairperson of COST269, 'User Aspects of ICTs'. She is the co-chair of the international association 'The Society for the Social Study of Mobile Communication' (SSSMC) which aims to facilitate the international advancement of cross-disciplinary mobile communication studies. Her work has been published in eleven languages.

Eric Harwit is Professor of Asian Studies and a lecturer in the College of Business at the University of Hawaii, and an adjunct fellow at the East-West Center in Honolulu. A 1984 graduate of Cornell University, he received a diploma from the University of International Business and Economics in Beijing in 1990, and his PhD in Political Science from the University of California at Berkeley in 1992. Professor Harwit's areas of expertise include the automobile and telecommunications sectors in China. He is the author of the books *China's Telecommunications Revolution* (Oxford, 2008) and *China's Automobile Industry* (M.E. Sharpe, 1995), and has written several other articles and book chapters on both the automobile and telecommunications industries. Professor Harwit is a frequent speaker at international conferences, and he and his work have been cited in *Time* magazine, *Businessweek*, *The Economist* and *Asian Wall Street Journal*, on MSNBC and National Public Radio, and in other print and broadcast media outlets.

Pui-lam Law received his PhD in sociology from the University of New South Wales, Australia. He is currently an assistant professor in the Department of Applied Social Sciences of The Hong Kong Polytechnic University and a visiting research fellow at the Centre for Creative Industries Studies of Peking University. His research interest is on modernity and social development in China. He co-authored (2004) *Marriage, Gender, and Sex in a Contemporary Chinese Village*, co-edited (2006) *New Technologies in Global Societies* and co-edited (2008) special issues on ICTs and China and on ICTs for *Knowledge, Technology and Policy* 21, issues 1 and 2, and is the editor of *New Connectivities in China* (forthcoming). Recently, he has been working on ICTs and the issue of migrant workers' identity in southern China.

Angel Mei-yi Lin received her PhD from the Ontario Institute for Studies in Education, University of Toronto, Canada, in 1996. She is currently an associate professor in the Faculty of Education, The University of Hong Kong. She has been an active and productive researcher in the areas of critical discourse analysis, urban ethnography, critical pedagogy, feminist media studies, and youth literacies and cultural studies. With a background in ethno-methodology, conversation analysis and social theory, her theoretical orientations are phenomenological, socio-cultural and critical. In recent years she has researched mobile SMS cultures of young people in Hong Kong and China, and has published in international research journals including the *Journal of Gender Studies*, *Discourse and Society* and *Continuum: Journal of Media and Cultural Studies*.

Yinni Peng obtained her BA (2002) and MA (2005) in Sociology from Peking University, and received her PhD in Sociology from the Chinese University of Hong Kong (2009). She was awarded a J. William Fulbright Scholarship in 2008–2009. She was a post-doctoral fellow in the Department of Sociology of Hong Kong Baptist University in 2009–2010. Currently, she is an assistant professor in Hong Kong Shue Yan University. Peng is widely interested in migration, labour politics and ICTs. Her journal articles and book chapters mainly focus on the social consequences of ICT usage in contemporary Mainland China. She co-authored 'The Use of Mobile Phones among Migrant Workers in Southern China' (*New Technologies in Global Societies*, World Scientific, 2006), 'Cellphones and the Social Lives of Migrant Workers in Southern China' (*The Social Construction and Usage of Communication Technologies*, The University of Philippines Press, 2007) and 'Mobile Networks: Migrant Workers in Southern China' (*Handbook of Mobile Communication Studies*, The MIT Press, 2008). She authored 'Internet Use of Migrant Workers in the Pearl River Delta' (*Journal of Knowledge, Technology & Policy*, Vol. 21, No. 1, 2008) and 'Exit: Power Struggles in a Time of Labour Shortage in South China' (*Social Transformations in Chinese Societies*, Vol. 6, 2010).

Jing Song gained a BA (2002) and MA (2005) in Sociology from Peking University and an MA (2007) and PhD (2011) in Sociology from Brown University. She is a post-doctoral fellow in the Department of Sociology, Hong Kong Baptist University. Her current research interests are market transition, stratification and mobility, political transformation, and family change, especially in post-reform China. Her published papers include 'Family and Market: Nonagricultural Employment in Rural China' (*Chinese Journal of Sociology*, 2010), 'Moving Purchase and Sitting Purchase: Housing Reform and Transition to Homeownership in Beijing' (*Housing Studies*, 2010), 'Daxuesheng de ziwodingwei yu shouji xiaofei' [College Students' Self-positioning and the Cell-phone

Consumption] (*Journal of Social Theory*, 2005) and 'Jingji tizhi biange yu cunzhuang gonggong quanwei de tuibian' [Economic Reform and the Transformation of Village Public Authority] (*Zhongguo shehui kexue* [Social Science in China], 2005). In 2007, Song won the Alden Speare, Jr. Memorial Award for the outstanding MA Thesis from Brown University.

Jing Wang obtained her MA in Sociology from Peking University in 2006. Her academic focus was migrant workers in the Pearl River Delta, the 'factory of the world'. She has conducted extensive fieldwork in the area, interviewing both the workers and their employers, and also touched on the issue of the spread and uses of the mobile phone there. She has also participated in research conducted by the Sociology Department, Peking University, in Beijing (on aging), and Sichuan and Shanxi (on how villages in China are changing). Upon graduation, she was recruited by SDX Joint Publishing House as an editor, which is renowned in China for its academic publications.

Boxu Yang received his PhD in Sociology from the University of Toronto, and is currently a professor at the School of Journalism and Communication, Peking University. He is the author of *Dianzi meiti de kuosan yu yingyong* [The Adoption and Post-Adoption of Electronic Media in Urban China] (Wuhan: Huazhong University of Science and Technology Press, 2001), *Quanqiuhua: qiyuan, fazhan he yingxiang* [Globalization: Its Origin, Process, and Development] (Beijing: People's Press, 2002) and *Yintewang yu shehui* [The Internet and Society] (Wuhan: Huazhong University of Science and Technology Press, 2003). His journal articles, mainly published in Chinese, deal with topics such as 'The Transition from International Communication to Global Communication', 'Social Networks and Local Communication', 'Public Space, Private Space, Personal Space and Communications in China' and 'The BBSs (Bulletin Board Systems) on the Chinese Internet and communication'. Current research includes the effect of BBS communication, SNS communication and social interaction on the Chinese Internet.

Shanhua Yang is currently a professor in the Department of Sociology at Peking University. He conducts field research in various areas of China, including Shanxi, Zhejiang, Hebei, Sichuan, Guangdong and Shanghai, and has published extensively on rural Chinese society and the lives of rural Chinese, including migrant workers. He co-authored *Chengxiang jiating – shichang jingji yu feinonghua beijing xia de bianqian* [Rural–Urban Families: Transition under the Background of the Market Economy and De-ruralization] (Zhejiang renmin chubanshe, 2000) and co-edited: *Dangdai zhongguo nongcun yanjiu (xia): shizheng daocha* [Studies on Contemporary Rural China, Vol. 2, Empirical Investigation] (World Scientific, 2000); *Nongcun jiceng zhengquan yunxing yu cunmin zizhi* [The Political Functioning and Autonomy of Rural Chinese] (Zhongguo shehui kexue chubanshe, 2001); *Dangdai zhongguo nongcun de shehui shenghuo* [The Social Lives of Contemporary Rural China] (Zhongguo shehui kexue chubanshe, 2005); and *New Technologies in Global Societies* (World Scientific, 2006).

Series editors' preface

The 'rise' of China comes with opportunities and challenges. The three-decade 'Open Door' reform has reshaped China's development path since the end of the 1970s. China has achieved impressive economic growth and it has now become a great world power which attracts and demands the attention of both developing and developed countries. Despite the impressive economic accomplishments, the magnitude of social problems in Chinese society has been exacerbated. Such problems include poverty, unemployment, migration, land disclosure, rural community, family disintegration, youth delinquency, HIV/AIDS, prostitution, mental illness, ethnic tensions, environmental pollution, and others. In addition to these problems, popular discontent with the government is also rising, which has led to violent clashes and rural unrest in recent years. Facing these challenges, the Chinese government has been attempting to alleviate the attendant social upheavals since 2003 by calling for 'people-oriented' social policies in order to 'deepen the reform of the social welfare system'. The government has also revitalized social work education and in 2006 established a system for the vocationalization and professionalization of social workers as a profession to foster social development.

Meanwhile, the 'rise' of China (Chinese mainland) has also impacted on social transitions in Hong Kong, Macao and Taiwan, not only in the political and economic field, but also in the social realm. Since the end of their colonial era in the late 1990s, and particularly in the wake of the two financial tsunamis, Hong Kong and Macao have established increasingly close political, economic and cultural ties with Mainland China. This has subsequently led to gradual changes in policies on tourism, education, immigration and social welfare, as well as the self-identity and sense of belonging in these two cities. Despite political uncertainties, Taiwan has also experienced similar changes in areas such as tourist and immigration policies and cross-border marriage.

This book series, 'Social Work, Social Policy and Social Development in Greater China', is designed to accommodate the growing need for disseminating new trends, practices and reflections on hot issues that have emerged as a consequence of social changes and social development in Mainland China, Hong Kong, Macao, and Taiwan from 1997 to the present. The series will explore and respond to the impacts of marketization, globalization, urbanization, modernization, migration, technicalization, consumerism, natural and social disasters on people's livelihoods and well-being in the Chinese societies. It comprises contributions from diverse disciplines, including social work, social policy and sociology.

This book series will attempt to fill this gap by first introducing how the reform has changed people's lives from different social aspects, such as the impact of globalization and urbanization on people in contemporary Chinese societies. Secondly, the series will examine the barriers and threats confronting people from urban and rural areas in enhancing livelihoods during this period of social transformation. And finally, the series will explore and discuss the impact of social work, social policies and social participation on social development in the long run.

Angelina Woon-ki Yuen-Tsang
Benjamin Hok-bun Ku
Ngai Pun

1 The great mobile revolution and the Chinese encounter

An introduction

Rodney Wai-chi Chu

This is an edited volume about mobile phone use on Chinese soil, with an emphasis on Mainland China. The 'China going cyber' phenomenon is becoming impossible to ignore in the era of the information society. How do the Chinese live with cyber China? How does the mobile phone as an information and communication technology (ICT) device work on Chinese soil? The 'living with' and 'working on' here must be understood in a dynamic manner – people and technology are interplaying with one another. Such emphasis implies that we do not adopt the impact–response framework as an analysis of the process, which could mean Chinese society is a passive receiver of the ICT era. Instead, we believe that, when the technology meets the culture, a kind of structurating process (in Giddensian language) takes place and they both shape and reshape each other. The two theses of the book, 'techno-social tension under a compressed modernity' and 'Chineseness of mobile phone use', are attempts to show how the social dynamics of China are interacting with the technology. And in the process, the Chinese cultural elements are transformed into ICT practice in light of the rapid uplift caused by the 'double juggernaut ride' elaborated below. We term the whole phenomenon the 'Great Mobile Revolution' not only because the mobile phone itself signifies a kind of 'fourth communication revolution' (Fortunati in this volume, Chapter 4), but also because it is bringing in a new echelon of life experience for the Chinese in terms of interaction, aspiration and life-goal formulation. However, the path of this endeavour is not an easy one.

To start with, social scientists in general, and particularly those focusing on the ICT field, have shown great interest in the ways ICT devices may bring about changes to the livelihood of the people, and mobile phone usage is definitely an aspect that has been drawing much attention. But as will be shown later in the book (Chu in this volume, Chapter 3), although mobile phone studies in the West have blossomed, in China, researchers have turned their attention more toward the Internet than the mobile phone. And within the mobile-phone-related endeavours, there appears to be a political sociology bias. In this way, we argue that the twists and turns within the cultural realm emerging and evolving from mobile phone usage can be overlooked, in particular when the ICT experience in China is never linear. The argument of this book is that, as a means of comprehending the rapidly developing Chinese society, the value of studying the mobile phone is no less important than the contribution that study of the Internet can offer. Fortunately, we have reached the right moment to make such a claim. There is now enough evidence to show that it has become difficult, if not impossible, to differentiate between mobile phone usage and Internet usage in light of the technological advancement in ICT. According to the report of the China Internet Network Information Centre (CNNIC 2009), by end of June 2009, with the launch of the 3G service, the number of mobile Internet users had reached 155 million. What is more, according to the survey mentioned in the report, 28 per cent of the mobile Internet users said they would access the internet by 3G mobile phone, and of those who had not adopted the 3G service, 49 per cent said they would like to use a 3G mobile phone to access the Internet. It is obvious that the boundary between mobile phone users and Internet users is blurred. For instance, Harwit's study has revealed that, by mid-2010, about 66 per cent of Internet users in China (nearly 280 million people) accessed the Internet via mobile phone (Harwit in this volume, Chapter 6). As such, one should not doubt the claim that the mobile phone as an ICT device is playing a significant role in cyber China.

The sociological meaning of the mobile phone in the current era

To take stock of the mobile phone literature produced in the last two decades is a daunting task, and there is strong evidence to show how the mobile phone has its own characteristics that not only older media, but even computers and the internet, do not have (Fortunati in this volume, Chapters 2 and 4). It is interesting to see how the attention given to the mobile phone can be associated closely with the scheme of Giddens' construct of his structuration theory (Giddens 1984). The first theme of the structuration theory is 'separation of time and space', which indicates that interaction among others is no longer constrained by the importance of presence, and so fosters relations between absent others (Giddens 1990: 19). A closely associated theme is the notion of 'dis-embedment', which refers to the lifting out of social relations from local contexts of interaction and their restructuring across indefinite spans of time–space (ibid.: 21). The third theme is the 'reflexive nature of modernity', or what Beck describes as 'reflexive modernization' (Beck 1992: 16), which means social practices are constantly examined and reformed in the light of incoming information about those very practices (Giddens 1990: 38).

Simply put, in the era of the information society, the meaning of time, space and physical presence has been redefined, where previous stable, bounded and institutionalized channels have become decentralized, and likely to be more individualized. In terms of the mobile phone research agenda, we have, for instance, Wellman's 'network individualism' to illustrate how the individual has become the primary unit of connectivity in the cyber age (Wellman 2001), Gergen's 'absence presence' to show the emerging significance of absent others in relationship building and maintenance (Gergen 2002), as well as the stress that the mobile phone is thus transforming contemporary cultural and social practices far more than fulfilling a communicative purpose (Goggin and Hjorth 2007). However, the mobile phone can also support tendencies toward social closure, as its use can be a way to hide from any specified time–space identification, thus presenting particular kinds of interpersonal and organizational uncertainties (Green 2003). What is more, besides empowering, it can also be seen as a means of controlling, as the possibility that 'you cannot be found' is functioning side by side with the possibility of 'forced accessibility' once you have got a mobile phone number (Kopomaa 2000; Puro 2002). The crucial point of concern is that the device itself is perhaps the most influential of late modern instruments in blurring the public and private spheres (Fortunati 2002), or what Puro contends as privatizing public space (Puro 2002). So much so that how can we incorporate this into the Chinese context?

The theses of compressed modernity and Chineseness

Mainland China has been experiencing a double-digit economic growth and rising per capita income from the late 1970s into the 1990s and on into the new millennium (Selden and Perry 2010) that has been the envy of the world. But this kind of economic prosperity has been accompanied by a characterization of deepening social rifts in terms of political, social and ethnic conflicts, and so in recent years the call for 'social stability' has become the most common phrase in the official rhetoric (Lee and Hsing 2009). As a matter of fact, recent China study literature has been trying to capture the dilemmas faced by those who are living through this era of rapid modernization, reflected in, for example, life chance, life choice and the struggle between 'right and good' (see, for instance, Chu and Ju 1993; Goldman 2005; Hertz 1998; Link *et al.* 2002; Lull 1991; Pun 2005; Walder 1986; Yuen *et al.* 2004; Zha 1995). But we want to argue these attempts are at most covering only one aspect of China's rapid development – the industrialization and/or the capitalization process. What makes China's modernization so compressed is that she has a second force in action. Our stress here is that ICT exposure has further accelerated the speed and scale of dynamic changes that the Chinese have had to face in the last decade, thus resulting in more twists and turns in the normative and behavioural aspects of the contemporary Chinese – no matter whether they are directly in touch with the cyber world or not. Thus, when we mention that the first thesis of the book is 'techno-social tension under a compressed modernity', we maintain that the 'tension' is not just a matter of distribution or redistribution of resources (even in the broadest sense) from a socio-economic perspective, but a kind of 'forced into' mood of experiencing a fast-changing process where the social fabric is still trying to adapt itself to the pace of capitalism.

Such a compressed mood can be articulated with an interesting example. As shown in Table 1.1 which captures the number of phone users nationwide in China, we can observe how land phone users are outnumbered by those of the mobile phone despite the fact that the former developed at a much earlier time. In many areas in China, just as in the case of other Eastern countries and in Latin

Table 1.1 Number of phone users nationwide (1991–2010)

Period (year end)	*Fixed phone*[a] *(incorporating both village and city lines)*	*Penetration rate*[b]	*Mobile phone*[a]	*Penetration rate*[b]
1991	NC[c]	1.3	5	NC
1992	NC	1.6	18	NC
1993	NC	2.2	64	NC
1994	NC	3.2	157	NC
1995	4,071	4.7	363	0.3
1996	5,495	6.3	685	0.6
1997	7,031	8.1	1,323	1.1
1998	8,742	10.5	2,386	1.9
1999	10,872	13.0	4,330	3.5
2000	14,512	20.1	8,453	6.8
2001	18,039	NC	14,522	11.4
2002	21,400	NC	20,662	16.2
2003	26,331	23.7	26,869	20.9
2004	31,224	24.9	33,482	25.9
2005	35,043	27.0	39,343	30.3
2006	36,781	28.1	46,108	35.3
2007	36,545	27.8	54,729	41.6
2008	34,080	25.8	64,123	48.5
2009	31,369	23.6	74,738	56.3
2010	29,438	22.1	85,900	64.4

Source: Ministry of Industry and Information Technology, The People's Republic of China, various years.

Notes

a Unit: 10,000 users, rounded up to the nearest digit.

b Unit: number of phones owned per 100 persons, rounded up to the nearest first decimal point.

c NC stands for 'not confirmed', because the early data of MIIT (those before the year 2000) were sometimes unclear in both categorization of figures and presentation style, but we believe the missing data should not affect the overall trend analysis.

America, the mobile phone is not an addition to the fixed line, but represents the user's first telephone line. Clearly, the number of phone users in China, in particular those using mobile phones, has rocketed in around two decades. From year end of 1995 to 2010, mobile phone users increased more than 230-fold; in June 2009, China formally, and finally, had overtaken the United States to become the nation with the highest number of mobile phone users in the world (CNNIC 2009). But still we are only talking of a penetration rate of around 65 per cent of the population, implying that there is still much room for expansion.

But our concern is of course not about figures. The table is just an example to capture the rapid modernization process in an extremely compressed fashion in the past 30 years, with growth rate in recent years being particularly speedy. Is this leap, and the social consequences of it, predictable and comparable? To elaborate, mobile telephony as a product of ICT is a comparatively recent phenomenon, but whereas its development as a communication tool in the first world is a fairly linear progress, the emergence and popularization of the device in China is more of a leapfrog because of the acute economic advancement of the country.

One may argue that this is normal if we look at other rapidly developing areas of the world, such as India, Russia and Brazil, that, together with China, are now known as the BRICs. But certain elements, namely her political system (a communist regime at least in name), demographic figures (more than one-fifth of the world's population lives in China), economic strength (especially since the financial crisis of late 2008) and role in international matters (Hu Jintao was ranked second on 11 November 2009 by Forbes in the 'The World's Most Powerful People' list, just after Barack Obama [see Forbes 2009]), when combined, may explain why China deserves special attention.

This thesis of tension caused by compressed modernity brings the second concern, the thesis of Chineseness, front of stage. Mobile phone use as a global phenomenon has a universal impact on people of the contemporary era, but is there a particular path of development under way in China due to the cultural and social context? When we look into the use of this ICT device in China, we can see behaviours and norms that differ from those found in the West. Here, we are talking about how network individualism, identity formation, relationship maintenance, the public/private divide, Short Message Service (SMS) usage, and the like as different aspects of mobile phone usage have to be understood in the Chinese context. Putting the two theses together, we can trace how they cut through the chapters of the book. For instance, media reform may generate social changes and have an impact on self-identity; but if the traditional Chinese social structure has not facilitated personal space or the necessary relational resources to deal with a social world that no longer has a fixed reference point, will the kind of 'network individualism' on Chinese soil bring about identity tension instead (B. Yang in this volume, Chapter 9)? On the other hand, as mentioned above, whereas in the West we find studies concerning mobile phone usage and its blurring of public and private domains, Chinese culture seems to celebrate situation-specific and relationship-specific differentiation, so such blurring would not be considered a problem (Peng and Chu in this volume, Chapter 13), which is fully explained by the concepts of 'differentiated mode of association' (*chaxugeju*) (Fei 1992) and 'situation-centredness' (Hsu 1981/1953). What is more, if we understand the socio-cultural backdrop of the new migrant workers (Chu and Yang 2006: 226–27), then we may more likely interpret a unique meaning of SMS usage for the Chinese (Lin in this volume, Chapter 12), and understand the reason why China is a global leader in SMS use, as captured in Table 3.1 of this volume.

In sum, the 'double juggernaut ride' is referring to how the society is exposed to a rapid and yet almost parallel process of industrialization *and* cyberization. The former is mainly highlighted by the capitalization process of the Chinese economy; the latter is referring to people's normative basis and behavioural norms that can be observed as a result of the computerization and digitalization of the society. We maintain that such *density* of experience (i.e. dealing with modern capitalism and cyberization at the same time) under a *compressed period* when the society has a thick *cultural heritage* of its own, and yet is still under a *dictating rule*, represents a unique situation in China. As an agenda, the riding of the double juggernaut thus reflects a complicated picture that should consist of not only how the East meets the West, but also how humans interplay with technology, and tradition coincides with modernity. The purpose of this book is to make use of mobile phone usage in China to articulate the normative basis of the Chinese in the midst of this double juggernaut ride.

Content outline

Part I: Overview: from global to China

In Part I, the intention is to provide the answer to the two questions posed at the start of this introduction regarding the meaningfulness of using mobile phone studies to understand the contemporary era, and the uniqueness of China (Chineseness) to global mobile phone studies. As one of the pioneers in studying mobile phone use in Europe from as early as the 1980s, Fortunati provides us with a detailed account of the emergence and development of mobile phone studies in connection with the globalization of human society. By recapitulating the shift from study of the fixed phone to the mobile phone, the author moves to outline how the mobile phone has become a 'new medium' with rocketing speed of diffusion – reaching more than 3 billion people in less than 20 years. One of the mobile phone's characteristics is the way it imposes a new rhythm (in contrast with the computer and the Internet) to hybridize itself as an ICT device with the human body, and this also explains how the mobile phone can be analysed as a technological artefact, making it a device not in itself, but as an object that may express a consumer's subjective taste. Needless to say, the incorporation of Internet access into mobile phone design has given this ICT product an even more significant weight within the social fabric of the information society. All in all, the mobile phone has become a very rich research topic in the Western world because social scientists see its power to penetrate different aspects of life, and in a sense, study of the mobile phone is a response to a core sociological enquiry of this period: the interplay between humans and technology. But Fortunati is also aware of the fact that the stock of mobile phone literature still has room for academic endeavour from a cultural point of view if one wants to have a more realistic overview of the practices of mobile phone use from a global perspective. Such concern from her prepares the way for Chu to open up the discussion in the third chapter.

Chapter 3 in fact lays the foundation for depicting the Chinese context for the chapters in Parts III and IV of the book. Chu contends that ICT research on China in general – and thus including mobile phone study – is more often a means to shed light on the 'socio-political dynamics within a communist regime in its path of modernization'. This is in contrast with the Western endeavours which see the study of the mobile phone, as 'a sociological enquiry of humans and technology', as an end in itself. The general over-riding concern of China's ICT research with a political sociology aspect is caused much by the curiosity regarding how a communist regime could work to modernize herself, thus overlooking two aspects that make up the uniqueness of China: the 'reform mood' is working under a 'rule by morality' thesis which has a cultural heritage of its own, and the 'double juggernaut ride' is an experience that her counterparts have not been so experienced with. The simultaneous processes of industrialization and cyberization have resulted in powerful social and cultural dynamics, and the ways Chinese people of different walks of life are living with this global phenomenon need detailed analyses, as seen in this volume. The chapter ends with a revisit of China mobile phone studies hitherto, and projects for future study in this research area.

Part II: Theory, history and comparison

In Part II, the four chapters bring forth the theoretical, historical and comparative backdrop to the Chinese setting. We begin by documenting the theoretical reasons for mobile phone studies within the ICT field (Fortunati, Chapter 4), followed by a detailed description of mobile phone development history in Mainland China (Wang and Cheng, Chapter 5). From there, first a Sino-Japanese (Harwit, Chapter 6), and then a Sino-Western (Fortunati and Law, Chapter 7), comparison of mobile phone development and phone use will be put forward.

Fortunati opens the discussion again by arguing that the advent of the mobile phone can justify speaking of a 'fourth revolution' in human communication. This claim emerges from the examination of the role of the mobile phone in ICT history. The mobile phone must be distinguished from other media for the following reasons: this technology demonstrates the fastest and largest global diffusion because of its ability to cross illiteracy, its lower cost and its adaptability; it is the first ICT device that can enhance an individualization process with an ease that other media cannot match; it opens up a new ontology of the human being by its increasing role in the construction of identity in general, and in the social presentation of the self in particular; and it enables new frontiers of convergence via matching voice, text and images (with the increasing popularity of the 3G setup), testifying to the meaning of fusion as the key principle of globalization to its very end. As such, Fortunati argues that it is sufficient to say that the mobile phone, as one of the many ICT media, should be given more credit and attention as far as its sociological meaning is concerned.

In the only chapter with a historical dimension in this volume, Wang and Cheng trace the history of the development of mobile phones in China in a fascinating and detailed way. What makes the mobile phone, regarded as an unobtainable luxury for the bulk of the population in the late 1980s, become an ICT device that has a penetration rate of 65 per cent by the end of 2010, already enough for the Chinese population to be the greatest number of mobile phone users (85,900 m) in the world? The two authors divide the development process into four stages: the luxury stage (1987–1993), the high-income user stage (1994–1998), the mass popularity stage (1999–2004), and finally the competitive and diversity stage (2005–2010). This chapter is interesting in that it explains how the mobile phone market of China, which was started with only three categories of users (businessmen, high-rank government officials and those in special forces), grew into one that is full of a mix of real brands and counterfeits and no-name knockoffs (also known as '*shanzhai*' or 'mountain fastness' mobile phones). They examine the main catalysts in the rapid expansion of mobile phone usage in China: improvements in the general living conditions, the reform of the telecommunications industry, the maturity of mobile communication networks, and the flourishing of domestic mobile phones. Their ability to capture the emergence, development and decline (in one particular case) of SMS, WAP (Wireless Application Protocol), MMS (Multimedia Messaging Service), mobile games and *Xiaolingtong* (a technique of connecting to the network of fixed lines by wireless technology) not only provides the necessary data for the more analytical chapters on the day-to-day use of mobile phones, but also makes this chapter essential for readers to have a general grasp of the reality of mobile phone use in Mainland China today.

As one of the two comparative efforts in this volume, Harwit assesses how the mobile Internet network may grow in China by tracing similar developments in China's East Asian neighbour, Japan. The Japanese system, pioneered by NTT DoCoMo and other corporations, has transformed many aspects of Japanese social, cultural and commercial life. Harwit depicts the ways Japanese use the mobile Internet, on a small screen, for basic communication, group interaction and mobile commerce, and examines the successes and failures of the Japanese mobile Internet since its introduction in 1999. From there, he moves on to examine the development of China's mobile Internet in light of the Japanese experience. Do the early stages of mobile phone Internet use in China mirror those of Japan? What similarities and differences can we expect as mobile Internet use penetrates the Chinese population? In particular, what lifestyle differences and consumer preferences will be the key in shaping the way Chinese citizens adopt 3G mobile phone data communication? Will there be different trajectories in urban areas of China (built in a pattern geographically similar to that of Japan) versus the less concentrated and less affluent rural parts of the nation? Harwit predicts that the more restrictive political system in China will limit the ways the mobile Internet can be used in the PRC, and such restrictions will shape 3G use in the early years of the technology. But he is alert to the lively arena of discussion of cultural, social, economic and non-sensitive political issues circulating in the cyber environment in China that may affect many aspects of Chinese life. He concludes that as China's political and social liberalization continues on a steady course, we may see some convergence in the ways mobile Internet use grows in both China and Japan.

Fortunati and Law set out their enquiry by the 'one-machine–one-individual' thesis surrounding mobile phone use. They argue that the thesis seems to witness the rising concern with the individualization process in Western countries because of the transformation of this device from being a mobile technology to a personal technology. But does individualization appear in China to be similar to or different from the phenomenon that has developed in Western countries? For sure, the overwhelming growth of the capitalist system in China, with the rapid development of innovation and technology that has gone with it and the development of wage relations, is developing the process of individualization. Yet, to what extent are Chinese social and cultural traditions deviating, modifying, blocking or activating this process? By examining the theoretical and historical framework in which the individualization process has developed in old industrialized countries, they pave the way to understanding the structural configurations and characteristics of this process which are intrinsic to it, and thus bring out the cultural side that may be re-semanticized by the Chinese. Fortunati and Law support the argument that the use of the mobile phone is one of many examples of the hybrid nature of the current capitalist process in China, where traditional social structures endure. In such a light, the double aspect of the mobile phone is highlighted: a potentially empowering technology but one which is also a tool for valorization with structural and social consequences in the process of individualization in China. Thus, whereas the encroachment of globalization and the development of the capitalist system in China are changing the country profoundly, this process is seen to be changed deeply by China in turn. In the end, readers have to ask: how much should we prepare ourselves to face a kind of Chinese modernity?

Part III: Techno-social interplay: cultural and identity tension

As previously discussed, a key approach in ICT studies on China is a kind of political economy endeavour, with attention placed on the socio-political implications on Chinese soil that come about as a consequence of the launch and penetration of mobile phone use. We try to widen such scope in Part III, which witnesses a gradual move from a more political-economy concern (Elaluf-Calderwood, Chapter 8) to the tension between technology and identity (B. Yang, Chapter 9; and Fortunati and S. Yang, Chapter 10) as derived from the diffusion of mobile phone use in a context that has strong cultural roots. Together these three chapters show the 'thesis of tension' that is in evidence as China travels along her unique path of development.

Elaluf-Calderwood picks up the thesis of the development of civil society to depict Chinese modernity in the ICT era. By employing Heidegger's concept of *Dasein* and using the 2003 Beijing demonstration against the Iraq War that was held in front of the US Embassy, she illustrates how a political concern can be elaborated even in a socially and culturally more restricted place because of its mix with new social networks. To the author, being in the world in a post-modern era is a concept extended to not only the real world but also the virtual world. Compared with the Internet, which can overwhelm the user through information overload, mobile networks – signified mostly by SMS messages via the mobile phone – are much more personal and thus trustworthy. The suitability of the mobile network over the

Internet to promote individual participation in social action, together with the demands of post-modern life that require individuals to link up with loose networks, should facilitate the growth of civil society. It is here that we see that the philosophical ideas of *being in the world* and *being with others* may aid in understanding the role of social networks in the life of an individual and the society one lives in.

B. Yang assesses the social implication of the mobile phone. He argues that mobile phone communication has become an indispensable or organic part of today's social structure, especially in urban China. The idea is that modernization and globalization demand social mobility. The Western experience indicates that mobility means not only the disintegration of marriage and community but also the formation of social networks. This often means individualism *with* self-identity anxiety. Nevertheless, when social networks become necessary for our daily lives, media that enable the process of networking across physical boundaries have to be introduced and employed. With such hindsight, Yang explains how the mobile phone differs from the fixed phone in terms of the roles it plays in the social transformation in respect to modernization and globalization. Yet, this has particular meaning in the Chinese context. He borrows Barry Wellman's notion of 'networked individualism' to bring out how the process of glocalization starts in China, and yet, due to the particular political-cultural backdrop, brings about 'social liberation without freedom' and thus brings identity tension to the urban Chinese. Yang believes that both modernization and globalization pose severe challenges to the traditional Chinese social structure. After all, the Chinese community is never a community in the Western sense, and the social implications of the mobile phone in regard to relational properties such as trust, uncertainty and self-identity may be of a different kind on Chinese soil.

Indeed, we are facing how the end users of the mobile phone perceive the meaning of the machine to them. Fortunati and S. Yang choose to use the 'identity' given to the mobile phone as an entry point to unfold the discussion. Again, the query starts with cultural sensitivity: is the kind of identity given to the mobile phone the same to the Chinese as that of their Western counterparts? By using qualitative data collected from students, professors and white-collar and migrant workers, the two authors open up a two-front exploration: how users in China co-construct the 'identity' of the mobile phone, and the social changes that occurred in relation to the adoption of this device, analysing at the same time attitudes, opinions, expectations and behaviours of mobile phone users. It seems there is a Chinese way of constructing the identity of the mobile phone through the socialization process, and this makes it different from both Western and Japanese experience.

Part IV: Exploring the Chineseness of mobile phone use

At this point, it is important to remember that we are talking about the use of the mobile phone by Chinese, and not about the use of the mobile phone by Chinese *as perceived by* Western observers. This makes the move to Part IV of the book particularly meaningful, as different authors here (Chapters 11 to 14) make sense of the mobile phone use of different categories of users. Readers now face squarely the 'thesis of Chineseness'. Although the study targets of these four chapters are not a representative sampling of the general populace, the variations as seen in the socio-demographic characteristics (college students, migrant workers, white-collars) and the covered topics (self-positioning in the ICT era, gender perceptions reflected in cyber space, cultural threads as articulated in ICT practices, as well as social ties captured in the network society) are fruitful in understanding the practicality of the daily cyber life of the Chinese.

One of the undeniable facts of the current mobile phone market in Mainland China is that college students as a category form a key user group. As such, how this group of people perceives the mobile phone has socio-economic significance. Based on a series of interviews with college students in Beijing, Song and S. Yang uncover aspects of the students' perception of mobile phone usage. What is involved here is how the users identify themselves via their use of the mobile phone, and the kind of socio-ecological context surrounding them that constructs and directs their usage. One of the interesting findings from the two authors is that mobile phone consumption turns out to have a restrained and rational look, because the student group is basically relying on financial support from family seniors for day-to-day living expenses. Yet, as the proportion of people who can become college students remains low, there is a sharp contrast with the data on migrant workers. Simply put, students do not need to prove their social status by owning luxury models of mobile phone. The two authors stress how the socio-ecological context of this grouping – their living on campus – may bring about a very different consumption pattern. The students can normally get access to different kinds of communication instruments, and so there is always a choice. Naturally then, such a close environment will also imply that they may easily influence one another in the selection of the model of phone and in usage patterns, and interpersonal relations are in turn strengthened and reconstructed throughout the process. The authors argue that, if we want to depict mobile phone usage and its social implication for college students in China, we should take the unique self-positioning of the students into account and set this articulation against campus life in China.

The chapter by Lin moves to the analysis of the mobile cultures of the migrant workers working in Southern China through a textual analysis of SMS messages and the interpretation of messages of their users. In her endeavour, Lin shows how the migrant workers in Southern China have made extensive use of mobile text messaging not only to gain the most up-to-date information about the job market but also to create a totally alternative social space in which different sexuality and romance norms and practices can be tried out. Yet, given constraints on their literacy level (with their limited schooling experience), the workers have to rely a great deal on the SMS manuals which are organized as lists of numbered short message texts placed under different topics (e.g. Humorous Jokes, True Sentiment Confessions, Witty Love Letters, etc.). Lin conducts a textual analysis of these manuals, and uncovers a range of romance and sexual ideologies which constitutes a supermarket of radically different norms and recipes for practices. With the emergence and rapid diffusion of such ICT devices, the paradoxical relationships between the two sexes are being constituted in a manner that may reinforce or dilute traditional images, norms and practices of gender roles and relations, thus showing how deeply the social fabric is being transformed and is evolving in the ICT era.

The chapter by Peng and Chu extends the coverage to white-collar workers. They study how Chinese culture and social context modulate mobile phone use in China, and thus their work is an explicit effort in tackling the 'Chineseness' of mobile phone use. The authors explore how Chinese society shapes mobile phone usage via two special features: one is the situation-specific and relationship-specific usage of the mobile phone in the public sphere; the other is face (*mianzi*) and the differentiated interpretations of the mobile phone within Chinese society. Their data show that Chinese usually do not use the public/private division as the norm to regulate their mobile phone usage, but instead evaluate whether the co-presences are acquaintances or strangers, the situation is formal or informal, and whether their social status is equal or not. The whole issue at stake is not a public/private concern, but appropriateness of one's behaviour toward the other concerned party – the guidelines must consider the cultural codes of the Chinese. As they contend at the end of the discussion, it does not mean that Chinese users do not share some aspects of their use of the mobile phone with their Western counterparts, but there is a need to emphasize the indigenous perspective with its own social context that is interplaying with the global phenomenon of mobile phone communication.

Similar to the previous study by Peng and Chu, this endeavour by B. Yang may be regarded as exploring the interplay between the public and private domains. Yang targets registered citizens in Beijing and tries to identify the content of social ties when the mobile phone is employed as a communication medium. However, unlike the previous three chapters in Part IV, both migrant workers and college students are deliberately excluded because the mobile identity of the former and the limited exposure of the latter may not serve the research purpose of comparing and contrasting the different kinds of social ties the informants are exposed to. Beijing is chosen because it is a place that has a 60 per cent Internet penetration rate and 99 per cent mobile phone adoption rate, and such a study of social ties is rarely explored in the urban China context. The mounting literature has shown that social relationships are changing with the emergence of the Internet and mobile phone, so the researcher focuses on how users with different socio-demographic characteristics differ from each other in terms of their social networking activities and mobile phone communication patterns. Yang brings out the concept of 'imagined privacy' which concerns privatizing public spaces by means of imagination, but this concept does not deal with social situations such as personalizing private spaces (for instance, using SMS at home) in order to maintain personal and extraneous ties that may not have anything to do with those co-presented at home. It is with the 'privatization of public spaces' and the 'personalization of private spaces' that 'networked individuals' appear to be emerging even in the case of China, though it may need further effort to depict the degree of Chineseness within this kind of networked individual.

Future intentions and acknowledgements

Even with this scope of study, this volume is far from covering most, not to mention all, aspects of mobile phone usage with Chinese characteristics. We do, however, hope that this book is timely in alerting China watchers to the rich elements from the cultural realm that it would be costly to neglect. We admit that much of the data involved in this volume has been obtained through convenience sampling, while comparative data across the provinces of the Land of the Dragon to show whether different levels of opening up may affect the diffusion of ICT culture have yet to come. We understand well that the 'double juggernaut ride' claim in explaining the uniqueness of Chinese mobile phone use (or even a more ambitious theme to explore the broader modernization process of China) demands much more data support. In fact, the four editors in this volume (with David K. Herold from The Hong Kong Polytechnic University), in collaboration with Professor Chonglin Shen of the Chinese Academy of Social Sciences, have just completed a large-scale survey to collect the data of 4,800 cases from six provinces/municipalities (Guangzhou, Shanghai, Dalian, Nanzhou, Naning and Chengdu), followed by 120 in-depth case interviews, in order to better capture the practice of mobile phone usage in China. The geographical location of the field sites takes into consideration the different degrees of economic development in Mainland China and the relevant level of modernization and cyberization on the spot, and the elaboration of the data to enrich the understanding of ICT practice in China is the next stage of work. Having said that, we know very well that our emphasis on compressed development in China means we are more vulnerable to the risk of having obsolete data. A short time interval between two fieldwork trips (say three to six months) could already mean too long a period to reflect the ICT practice of our study target, if not the phenomenon itself. No matter whether we are talking about new migrant workers or young adults, some of our study targets may move so fast in their own life experience (the cyber one in this context) that, by the time the academics publish their findings, the findings are out of date and have become a *previous experience* (Chu 2008). It is an endless race, but this also explains why we must have a theoretical backdrop, a historical touch, a contextual grasp and a reflective attitude in light of this juggernaut ride through contemporary China.

A number of chapters in this volume (Chapters 5–6, 8–13) made their first appearance at the conference Mobile Communication and Asian Modernities II: Information, Communications Tools & Social Changes in Asia, which was held in Beijing in November 2005 and organized by the Research & Development Office of France Telecom in Beijing, China. The conference was a golden opportunity for scholars who share a sociological concern with the impact of ICT on human activities to come together. Needless to say, these chapters in their present form have been much extended and modified from the original conference drafts so as to fit (because of the logic explained above) with the current endeavour. Last but not the least, this volume could not have been published had it not been for the funding provided by the Department of Applied Social Sciences of The Hong Kong Polytechnic University. Special thanks must be given to Professor Angelina Yuen who, as HoD in the period 2004–2009, has supported and witnessed the emergence and development of the research agenda on ICT.

References

Beck, U. (1992) *Risk Society: towards a new modernity*, London: Sage.

Chu, G. and Ju, Y. (1993) *The Great Wall in Ruins: communication and cultural change in China*, Albany: State University of New York Press.

Chu, W.C.R. (2008) 'The dynamics of cyber China: the characteristics of Chinese ICT use', *Knowledge, Technology and Policy*, 21(1): 29–35.

Chu, W.C.R. and Yang, S. (2006) 'Mobile phones and new migrant workers in a south China village: an initial analysis of the interplay between the "social" and "technological"', in P. Law, L. Fortunati and S. Yang (eds) *New Technologies in Global Societies*, Hackensack, NJ: World Scientific.

CNNIC (2009) *CNNIC Publishes 24th Statistical Report on Internet Development in China*. Online. Available: www.cnnic.net.cn/html/Dir/2009/07/28/5644.htm (accessed 15 August 2010).

Fei, X. (1992) *From the Soil: the foundations of Chinese society*, Berkeley: University of California Press.

Forbes (2009) *The World's Most Powerful People No. 2: Hu Jintao*. Online. Available: www.forbes.com/lists/2009/20/power-09_Hu-Jintao_EWJW.html (accessed 16 November 2009).

Fortunati, L. (2002) 'The mobile phone: towards new categories and social relations', *Information, Communication & Society*, 5(4): 513–28.

Gergen, K.J. (2002) 'The challenge of absent presence', in J.E. Katz and M. Aakhus (eds) *Perpetual Contact: mobile communication, private talk, public performance*, Cambridge: Cambridge University Press.

Giddens, A. (1984) *The Constitution of Society: outline of the theory of structuration*, Cambridge: Polity Press.

—— (1990) *The Consequences of Modernity*, Cambridge: Polity Press.

Goggin, G. and Hjorth, L. (2007) 'Introduction', in G. Goggin and L. Hjorth (eds) *Mobile Media 2007: proceedings of an international conference on social and cultural aspects of mobile phones, media and wireless technologies*, Sydney: University of Sydney.

Goldman, M. (2005) *From Comrade to Citizen: the struggle for political rights in China*, Cambridge, MA: Harvard University Press.

Green, N. (2003) 'Community redefined: privacy and accountability', in K. Nyiri (ed.) *Mobile Communication: essays on cognition and community*, Vienna: Passagen Verlag.

Hertz, E. (1998) *The Trading Crowd: an ethnography of the Shanghai stock market*, Cambridge: Cambridge University Press.

Hsu, F.L.K. (1981/1953) *American & Chinese: passage to differences*, 3rd edn, Honolulu: University of Hawaii Press.

Kopomaa, T. (2000) *The City in Your Pocket: birth of the mobile information society*, Helsinki: University Press Finland.

Lee, C.K. and Hsing, Y.T. (2009) 'Social activism in China: agency and possibility', in C.K. Lee and Y.T. Hsing (eds) *Reclaiming Chinese Society: the new social activism*, London: Routledge.

Link, P., Madsen, R.P. and Pickowicz, P.G. (2002) *Popular China: unofficial culture in a globalizing society*, Lanham, Boulder, New York: Rowman & Littlefield.

Lull, J. (1991) *China Turned On: television, reform, and resistance*, London and New York: Routledge.

Pun, N. (2005) *Made in China: women factory workers in a global workplace*, Durham, NC: Duke University Press.

Puro, J.P. (2002) 'Finland: a mobile culture', in J.E. Katz and M. Aakhus (eds) *Perpetual Contact: mobile communication, private talk, public performance*, Cambridge: Cambridge University Press.

Selden, M. and Perry, E.J. (2010) 'Introduction: reform, conflict and resistance in contemporary China', in E.J. Perry and M. Selden (eds) *Chinese Society: change, conflict and resistance*, 3rd edn, London: Routledge.

Walder, A.G. (1986) *Communist Neo-traditionalism: work and authority in Chinese industry*, Berkeley, CA: University of California Press.

Wellman, B. (2001) 'Physical place and cyber place: the rise of personalized networking', *International Journal of Urban and Regional Research*, 25: 227–52.

Yuen, S.P., Law, P.L. and Ho, Y.Y. (2004) *Marriage, Gender and Sex in a Contemporary Chinese Village*, Armonk, NY: M.E. Sharpe.

Zha, J. (1995) *China Pop: how soap operas, tabloids, and bestsellers are transforming a culture*, New York: The New Press.

Part I

Overview

From global to China

2 From West to East and beyond

Main tenets in the studies of the mobile phone

Leopoldina Fortunati

Mobile phone study as a sociological endeavour

The framework in which the mobile phone is analysed nowadays is globalization (Castells *et al.* 2007). It is a matter of fact that the process of negotiation between societies and technology takes place in a global world which is experiencing changes in the organization of everyday life as well as in the conceptualization, production, and use of the mobile phone. Therefore, it is important to explore both how societies negotiate the amount of mobile phones that they need and/or which they in any case adopt, as well as their characteristics, and how the mobile phone does not elude social will, expectations, and requests (Chu 2008; Law *et al.* 2006; Oudshoorn and Pinch 2003, 2008). Like any other technology, the negotiation of the mobile phone has to come to terms with taken-for-granted practices; its penetration has to find cracks in the organization of daily life, it has to insinuate itself in new ways in times and places shaped by relations, sentiments, and emotions, it must meet with desires, needs, mistrust, or enthusiasm. But in its turn, the penetration of the mobile phone has made more acceptable social processes of extreme relevance, such as migration, short-range mobility, the transformation of the organization of labour in a post-Fordist sense, the formation of virtual communities of various kinds, and the maintenance of a link between the world of labour and the domestic sphere (Haddon 2004; Ling 2004). If it is true that mobile phone design brings with it user design, it is just as true that mobile phone users are increasingly able to invent functions and services and then to dictate future developments of this service (Pertierra 2007).

But in this framework, what is the meaning of the mobile phone at a social level? Defining the mobile as an instrument of communication is to say what it is for, and maybe redesigned for by its users. But this does not tell us anything about the meaning of its essential role, at the economic level, as a machine used within the social processes governing everyday life. The answer to this question is possible only if we connect the analysis of the role of the mobile phone (a) to the globalization logic and (b) the theory of domesticity (Fortunati 2006). With regard to globalization, the role of the mobile phone should be embedded in how global societies grasp the fundamental tendency of globalization, which is the principle of fusion. This principle describes the strengthening of the cooperation of labour and integration between the spheres of production and domestic reproduction and emerges as the structural logic of globalization. As for the theory of domesticity, applying it to the mobile in the reproduction sphere challenges in a certain sense the analysis of 'moral economy of the household' outlined by Silverstone (Silverstone 2006; Silverstone and Haddon 1996; Silverstone and Hirsch 1992). With this term, Silverstone refers to the process by which households appropriate commodities, inscribing in them specific meanings and transforming them into symbolic objects which contribute to the construction of household values (Ward 2006). If we connect the above analyses to that of Fortunati's on reproduction (Fortunati 1981), we can properly understand that the mobile phone is a *work tool for reproduction*. And as reproduction is extending its dominion over society today, the mobile phone can be considered a *strategic tool of social labour in general*. This analysis is supported also by the discourse carried out by Schütz and Luckmann (2003) who, as Gebhardt points out (2010), see technologies as *working tools* of everyday life.

If we continue to focus our attention on the process of diffusion and appropriation of the mobile phone, we need to elaborate a development strategy on the mobile phone taking the knowledge of global users, their attitudes, opinions, and behaviour as a starting point. The mobile phone as cultural object (Goggin 2006; Goggin and Hjorth 2009) is undergoing a re-semanticization process by local users, who, adapting and transforming its meaning, might generate the invention of new services and functions. At this point, the mobile phone should become flexible in its design so as to incorporate the mentalities of different people and cultures, languages, ethnic groups, genders, and generations. It is well known that globalization of markets is never a slavish imitation of the first phase of diffusion processes when it takes place at a second stage and in countries different from those in which this diffusion first appeared. As an example of this we can take the fixed telephone. In many areas of China, as well of other Eastern countries or Latin America, the mobile phone has not been an addition to the fixed telephone, but has represented the first and only telephone line. Innovation and the new have a power of attraction that is so strong as to make the re-proposal of any historicizing process impracticable. Global development proceeds in leaps, and has no difficulty in leaping immediately into the most recent innovation.

Studies of the mobile and fixed phone: a recapitulation

If we examine the studies carried out on the fixed telephone (see, for instance, Claisse 1988; Claisse and Rowe 1993; Claisse *et al.* 1985; De Sola Pool 1977; Dordick and LaRose 1992; Fortunati 1995; Katz 1999; Lange 1991; Lange *et al.* 1989, 1990; Moyal 1992; Rakow 1992; Rutter 1987) and those carried out on the mobile phone, we are immediately struck by two facts: the discontinuity between the research studies on the two types of phone, and the disappearance of research on the fixed phone. This discontinuity makes it difficult to trace the current use both of the fixed telephone and of the mobile phone and prevents the research and the patrimony of knowledge accumulated in the fixed-phone studies from merging into the field of studies on the mobile phone. In effect, this discontinuity tells us two important meta-elements: first, behind an apparent continuity from the fixed phone to the mobile phone (although the latter can be considered the 'child' of the former), these two tools have produced different narratives of reality and society; second, the comparative research between fixed-telephone communication and body-to-body communication, which was the main theoretical framework in which the studies on the fixed telephone had developed, has proved less of an attraction to those who were doing research on the mobile phone. For sure, had there been a kind of continuity between the lines of research on the fixed and mobile phone, it would have been valuable in the setting up of the field and understanding of the differences and similarities between the two artefacts and their practices of use. What has provoked this discontinuity?

A first reason can be summarized in a word: fashion. Following the advent of the mobile phone, the fixed phone was perceived as an object of research that was out of fashion. Second, there have been times when telecoms in a monopolist regime have been generous in funding research on the fixed telephone; but this generosity faded away when mobile phone sales provided most of the revenue. Third, a series of issues that have been discussed in relation to the fixed telephone have lost their attraction. For instance, while studies on the fixed phone focused on the comparison of telephone communication with body-to-body communication, and pointed out differences and similarities between these two different modalities of communication, the studies on the mobile phone do not replicate this kind of research. At that time, such comparison generated by fixed-telephone studies was necessary to understand what was happening to the world of communication in co-presence after an artificial transformer intruded into everyday life. Researchers investigated the relationship between the artificial and natural dimensions, between the communication in co-presence and that which mediated it, in order to understand the logic of choices among the different possibilities in the communication domain. With the mobile phone the researchers are more or less aware that body-to-body communication is not the prototype for every kind of communication. The fixed telephone has served in reality as a general experimentation in the world of artificial communication, in respect to which the mobile phone has little to add. Consequently, the analysis of the voice (Bologna 1992; Salazar 1987) as signal and infrastructure of body-to-body and telephone communication disappeared from the studies on the mobile phone and with it the problem of the lack of nonverbal language, with exceptions, such as Schegloff (2002).

The fourth element of discontinuity in the two tracks of research is the level of emphasis that scholars place on the mobile phone as technological and cultural object, which is lacking in fixed-phone studies. This discontinuity may have practical reasons since the fixed telephone is a very simple device with few functions and services such that over the course of time it became an invisible tool. In fact, the mobile is a more sophisticated technological artefact and its material body began to matter almost from the beginning (Glotz *et al.* 2005). For example, it has imposed the new issue of fashion, which was almost non-existent with the fixed telephone. Other new important issues are mobility and the body with the themes of hybridization, but also the dynamic between orality/writing, the call and the text. Further aspects that distinguish the research on the fixed telephone and the mobile phone are the following: (a) while the fixed telephone has allowed people to redefine and renegotiate the modalities of interpersonal communication, especially in the family, the mobile phone has allowed users to renegotiate the norms governing social behaviour in the public space and interpersonal behaviour in the intimate sphere of individuals; (b) while the former has been studied in detail by organization and business scholars in the work setting on the basis of its capacity to modernize an economy, the mobile phone has attracted much less interest toward its use in enterprises, with only a few exceptions (Ling 2004; Sørensen 2006; Sørensen and Pica 2005; Wiredu and Sørensen 2006); (c) the diffusion of innovation approach developed by Rogers (2003) has often been applied in the studies on the mobile phone, perhaps because the theory was contemporary to the diffusion of the mobile; in contrast, the studies on the fixed phone took for granted the issue of diffusion because this had already happened years before; (d) while the research on the fixed telephone reported very often on the use of this device for sexual harassment, with the mobile phone this thematic disappeared. Instead a common aspect in the two clusters of studies is the redefinition of space and time categories, although the mobile phone has explored new dimensions (Licoppe and Smoreda 2005). As for the debate on intimacy (De Gournay 2002), control (Haddon 1994, 1998), democracy (De Sola Pool 1984), and the dynamics of public/private spaces, they are common to the two debates, although mobile-phone studies present the analyses in these fields with a more sophisticated tone.

To recapitulate, the research issues investigated in the studies on the fixed and mobile phone only in part overlap. Contemporary China in this sense is a fascinating laboratory where a huge socio-technical system of mediated communication, especially mobile but also fixed telephone, has been set up in a very short period of time, and with a dynamic which sees the mobile phone diffusing more rapidly than the fixed phone (Law and Chu 2008a, 2008b).

Highlights of studies

Taking stock of the mobile-phone literature produced in the last two decades is an arduous task. The literature is made up of quantitative and qualitative studies mainly focusing on describing various practices of use of this device and the multiple dimensions – be it rational, emotional, or symbolic – that mobile phone use puts in motion in different social and cultural contexts. One can cite the review carried out by Geser (2004) with the purpose of structuring the field of sociological studies on the mobile phone, and a series of collected editions which include very interesting papers outlined in this framework (Hartmann *et al.* 2008; Höflich and Gebhardt 2005; Höflich and Hartmann 2006; Katz 2003; Katz and Aakhus

2002; Ling and Pedersen 2005; Nyíri 2003, 2005, 2008; Pertierra 2007; Pertierra *et al.* 2002). However, these studies include contributions coming mainly from Europe and North America, with some from Asia, while there are few contributions from Africa and South America. So from a cultural point of view there are still gaps to be filled in order to have an overview of the practices of use of the mobile phone from a global perspective. Another gap which can be observed is related to a distortion of the data at the researchers' disposal, which mainly concentrated on adolescents and youths with more limited attention paid to children, the elderly, or more mature users. A further gap is that there is little research carried out into women's use and practices of use of the mobile phone, with few exceptions (Hjorth 2005; Rakow and Navarro 1993; Wajcman *et al.* 2009).

The description of the practices of use of the mobile phone in situated contexts and temporalities would usually be the first step to building on theories. This kind of research is necessary to know what we are talking about, with subsequent researchers addressing crucial social issues on the basis of the knowledge acquired by exploring the structural dimensions of mobile phone use. However, looking at the whole research carried out on the mobile phone, Sawhney points out that researchers generally tend to focus on descriptions, and rarely bring them back to social problems or themes (Sawhney 2007). Furthermore,

> they tend to look at things through the technology lens – who (gender, age, socioeconomic, psychographic, and other such variables) uses the new technologies and why (social interaction, emergency, fashion statement, addiction, and so on). What is the content of the communication (content analysis, socio-linguistic analysis, and so on)? How do the new technologies change X (X = politics, workplace, education, and so on)? How does the development of the new technology differ in country Y1 and Y2 (Y = the US, Britain, Korea, and so on)?
>
> (Sawhney 2007: 396)

In addition to the cited theory of domesticity mentioned before, in which Fortunati has tried to understand the social meaning of the use of the mobile phone by looking at this device as a work tool conveying a large amount of immaterial labour (Fortunati 2006), domestication theory was outlined by Silverstone (Silverstone and Haddon 1996; Silverstone and Hirsch 1992). To depict the integration of ICTs – mobile phones included – in everyday life, Silverstone distinguished four stages of domestication: commodification, objectivization, incorporation, and conversion of technologies, seeing domestication as a process of consumption 'in which consumption was linked to invention and design, and to the public framing of technologies as symbolic objects of value and desire' (Silverstone 2006: 232). The application of the theory on social representations to the integration process of the mobile phone in the domestic sphere deserves a discourse apart. The theory on social representations gives more psycho-sociological concreteness to the domestication theory and has offered much data to understand how the mobile phone positions itself in cognitive maps and in the network of relationships which wrap them. People build their representation of a technology at the beginning of its diffusion and especially when, as in the case of communication and information technology, two opposite positions arise in the public debate. These cognitive processes must be monitored because they outline the assimilation and accommodation which occur in the stage in which we integrate cognitively a new object in the social thinking. Finally, Richard Ling analyses the mobile phone as a tool to strengthen social bonds among family and friends and to enhance social cohesion. He analyses how mobile communication is embedded in social rituals and how in turn it creates new rituals (Ling 2008).

Whereas Castells and others offer the broadest picture of the research on mobile phones in the developing world addressing the question of globalization (Castells *et al.* 2007), Jonathan Donner in an important review of the studies carried out in the developing world distinguishes two areas of the analysis: the first, made up of works on the technologies and e-actors; the second area, focused on studies with a strong economic development perspective (Donner 2008). The work of Boczkowski and Lievrouw (2008) individuate three tendencies in the literature on new media which include the mobile phone: (a) causality in technology–society relationships, which actually coincides with the last but one track proposed by Sawhney (2007); (b) the technology development process to better understand which of the theories of domesticity, domestication, and social representation have been advanced; and finally (c) the social consequences of technological change. We propose to add to this list another two topics: (d) time and space, which co-shape the social reality of each medium, the mobile phone included, and whose reformulation is in the background of much of the research and discourses developed in the studies of the mobile phone; (e) convenience, a driving principle which guides many of the social behaviours expressed in the domain of the use of the new media. Under the apparent rational choice on the use of one device or of another, often the principle which is at work is that of convenience, otherwise called Occam's law.

The mobile phone as a new medium

The first point to understand is to what extent the mobile phone is a new medium. Does it represent an evolution of the old media or a true revolution? Van Dijk (1999) analyses this issue, recalling that Williams (1982) has identified the second concentration of innovation at the end of the twentieth and beginning of the twenty-first centuries. It seems we can speak of a revolution, but must keep in mind that a revolution – even if it represents a moment of discontinuity, a jump in respect to the past – has also elements of continuity with the past and also elements of regression, not necessarily of positive progress and improvement. A follow-up question is whether the mobile phone should still be considered a new medium after more than one decade of its diffusion and appropriation. The mobile phone is often analysed as a new medium, although the notion of 'new media' is risky because it is based on a concept of novelty which is in itself a temporary concept (Kim and Sawnhey 2002). The answer is maybe problematic. So let us put aside this question and ask what the features and capabilities that characterize it are. We might consider a series of characteristics that come out from a classification exercise (Star and Bowker 2002). The mobile phone can be defined on the basis of one of its physical characteristics, such as its being a mobile or a portable or even a wearable technology, but which in time has happened to be used also in a sedentary way. Or it can be defined by the structure of the messages that are conveyed through it, structure which is generally circular, bi-directional, and both synchronic (calls) and asynchronic (Short Message Service, SMS), but which can also be made up of one-way direction messages (SMS news, horoscope, weather forecasting, and so on) and even one-to-many direction messages (the SMS messages that are sent to a list of friends, for example, on the occasion of a festival). Furthermore, the definition of the mobile phone might be linked to the modality of its use, which is largely personal in industrialized countries, but which in many developing countries can be collective (Donner 2008; Pertierra *et al.* 2002). Another way to define the mobile phone is through the characteristics of the language conveyed by it, which are multiple. The language can be in fact oral, written, visual, audio (sounds or music), and mixed. Or the mobile phone can be defined according to the dimension of the infrastructure it requires, which is, according to Schramm's definition (1977), that of the big media, since mobile phone systems require expensive, extensive, and complex infrastructures. But apart from this latter one, these features have become in time spread features of the mobile phone, after an intense negotiation with the second and third waves of users. In fact, the mobile phone has been completely reshaped by e-actors (Fortunati *et al.* 2010; Gebhardt *et al.* 2010).

As for the characteristics of the mobile phone as a 'new medium', it is more interactive, more multimedia, more integrated, and more flexible in comparison to the fixed phone. Moreover, following Manovich's suggestion (2001), it is worth looking at the mobile phone as a medium born of the tradition of communication in real time. The main point of tele-presence is in its anti-presence: it is no longer necessary to be physically present in a certain place to affect the surrounding reality. In this situation the distance between the subject and the object or other subjects becomes the crucial point, since it is the distance which shapes perception. According to Virilio (1992), small optics implies the distinction between near and far, between an object and the horizon, while big optics refers to the electronic transmission of information which eliminates the distance, and to the vastness of space, which implies a temporal difference between events and our reactions. The mobile phone, which represents the mobile part of telecommunications, becomes the quintessence of post-modernity, if we consider this latter as the process that eradicates the obstacles to circulation. The mobile phone is the technology which comes closer to the human body and which reduces the distance from the body. This has many implications for processes such as the hybridization between the body and the technology, and the change of direction which its development has inevitably led to, for the big family of information and communication technologies.

Up to now the technological process has always represented the extroflection of the communicative or work capacities of human beings from human beings themselves toward the environment surrounding them. The distance between devices and users was a great stimulus for social and personal reflexivity in technology. The new media invert this tendency. Already other media such as the radio and Walkman had reduced this distance from the human body, but these were not in a certain sense mainstream technologies. The computer and the Internet were the first to reduce the distance of the human body from the screen, but the mobile phone has imposed a new rhythm in this escalation, arriving to hybridize itself with the human body. Worn on the human body, the mobile phone eliminates the distance between the subject who sees and the object which is seen and puts in crisis the perception (which needs distance). Given that it is the distance that allows the objectivization, the mobile phone reduces its possibility.

The mobile phone as a technological artefact

The mobile phone has also been analysed as a technological artefact. The reasons for e-actors to reshape the mobile phone are connected to a plurality of factors: the economic factor is important, but also the aesthetical and the convenience factors. E-actors enthusiastically adopt, for example, SMS, which at the beginning was free and therefore much more convenient than calls. On the basis of aesthetical reasons, the mobile phone was obliged to become a fashion object, an ornament, or even a jewel. At the same time a series of design variants on the material body of the mobile phones were introduced by e-actors (Haddon *et al.* 2005).

Fortunati (2005) traced a map of the design variants registered in the current literature, which we briefly summarize here: (a) the individualized character, (b) the sedentary use, (c) its overtaking the landline phone, (d) a new dynamic between the private and public dimensions, (e) its attracting in the private dimension costs from work-communicative traffic, (f) the use of text messages, (g) the use of special ring signals (or beeping), (h) the practice of calling random numbers in order to widen one's social circle, again introduced by adolescents, (i) the practice of interacting with the media, (j) the use of the mobile phone to present oneself, (k) its aesthetic aspect, and (l) the attribution of great importance to music on the mobile phone.

These user modifications of the mobile phone should be seen as the expression of consumer subjectivity and innovation. E-actors begin to see themselves as an independent variable of the economic process, trying to cut costs and to adapt what they have bought to their needs and desires, thereby avoiding the command scheme embodied in every device and applying all the modifications they consider suitable (the famous 'dystopias of consumption'). In this framework, considering e-actors as an independent variable of the process of production of mediated communication makes intelligible and meaningful the effort carried out by researchers in assuming a contextual perspective capable of investigating local and everyday practices (Oudshoorn and Pinch 2003, 2008). The mobile phone has also been analysed in some studies as a dispositif, in the revisitation that Agamben (2009) produced of this Foucauldian notion (Lasen 2010). Many studies, moreover, address the question of its being a cultural (Goggin 2006, 2008), emotional (Vincent 2003; Vincent and Fortunati 2009), and symbolic object (Yung 2005). The mobile phone, just because it is an object of boundary on the merging of powerful dimensions, is able to reveal unexpected modalities of our current relationship with technological artefacts.

The mobile phone vs. mobile communication

To carry out this excursus we chose to look at the mobile phone rather than at mobile communication, because if we had considered mobile communication, we would have risked operating a reduction of the multiple factors implied in the social process developing around this technology. The danger would have been to not acknowledge all the implications of the mobile phone as a technological artefact and all the tasks and services which are conveyed through the mobile phone. In addition to conversation and information, the mobile phone in fact conveys also orientation, localization, agenda, time measuring (clock) and alerting (alarm clock), mobile business, entertainment, m-learning, and so on.

A structural analysis of the specific characteristics of mobile communication compared to telephone communication and body-to-body communication is still lacking. Höflich has made much effort in this direction (Höflich and Gebhardt 2005; Höflich and Hartmann 2006), but a substantial study capable of challenging the theories of mediated communication regarding the mobile phone is still to be done. Here it would be necessary to contrast mobile communication with telephone communication and body-to-body communication, but the main difficulty is represented by the discontinuity of the research on fixed and mobile communication carried out up to now. A robust attempt to investigate the ontological pre-conditions of mediated interpersonal communication from an e-actor point of view has been carried out by Gebhardt (2010). By focusing on inter-subjectivity, understood as the problem of co-ordinating reciprocal communicative actions, Gebhardt returns to the socio-phenomenological concept of the ‘everyday life-world’ as it has been elaborated by Alfred Schütz, and later by Schütz and Luckmann (2003). He proposes to conceptualize communication technologies as communicative working tools, whose interpersonal use is based on reciprocal orientations, rules, and mutually established social practices. But we believe that it does not matter as much what one chooses as the main focus as how one frames it. Starting from each point we can reconstruct the processes we are interested in examining. In fact, the theme of mobile communication crosses many of the studies produced in this field.

The mobile phone as a socio-technical system

The mobile phone should be examined as a socio-technical system (Bijker and Law 1992; Bijker *et al.* 1987), since this notion can include most aspects considered up to now. As a socio-technical system the mobile phone contributes, with all the other socio-technical systems built around the other media, to the huge socio-technical system which forms the communicative environment in which we live, and which enhances, limits, or complements body-to-body communication. If we take into account that technologies are the infrastructures of the communicative environment and interact in our everyday life setting, the relationship with each technology, the mobile phone included, is embedded in a layer of networked relationships which build meanings, symbols, metaphors, dreams, attitudes, opinions, emotions, and behaviour (be it conflicts, resistance, or appropriation), thus making life a huge socio-technical system. In a given moment, this global socio-technical system, which includes all the technologies diffused in the workplace, in the public sphere as well as in the domestic sphere, interacts with other empires of signs and families of objects. Analysing the mobile phone in this way becomes relevant not only because e-actors are used to returning to a cluster of technologies in their everyday life, but also because of the convergence among technologies and not only technologies. There is in fact the tendency to create convergence between technologies and objects belonging to other families such as textiles or biological elements and even a convergence between technology and social phenomena such as mobility, and so on.

However, the repertoires of the research in the telephony domain present, as we have already stated, a problem of discontinuity of their tenets and so also a problem of accuracy in the understanding of the mobile socio-technical system. In fact we have insufficient data on the recent evolution of the telephony: for example, why does video-telephony go so slowly? How do the fixed phone and the mobile phone converge? Which future will mobile media have? These questions remain open, even if among the various socio-technical systems the mobile phone (with Internet access) represents the most important innovation in respect to the past. In fact if we look at the new media from a user's point of view, the mobile phone shows the highest speed in its diffusion, reaching more than 3 billion people in less than 20 years. This means that all around the world people find this device convenient, friendly, and easy to use. But what is more important is that people, deciding worldwide to privilege the mobile phone in respect to the fixed telephone, also focus at the same time on reinforcing body-to-body conversation through both voice and writing.

If we follow the analysis we have carried out hitherto, an ideal study would be one that centred on all technologies at the same time. But the operation of such a study would be so complicated that this idea is by default too difficult to carry out. Alternatively, we can study also a single technology, as in the case of this book – the mobile phone – but being aware of the fact that we are producing a simplification of a complex problem and being clear at the end that what we have offered is an operation of simplification. After all, the problem is not in simplifying, since this has proved many times to be a very efficient strategy, but in not acknowledging that we are simplifying.

The previous pages, though sweeping, have shown the abundance of literature surrounding mobile phone study in the West. Yet, to those who have followed mobile phone studies on China, it will be clear that, in China studies, the path and foci of concern of academic endeavour is different. The cause of such a difference must be dealt with, and we maintain that it has to be understood in a wider context, which is the intention of the next chapter.

References

Agamben, G. (2009) *What is an Apparatus? And other essays*, trans. by D. Kishik and S. Pedatella, Stanford, CA: Stanford University Press.

Bijker, W. and Law, J. (1992) *Shaping Technology/Building Society: studies in socio-technical change*, Cambridge, MA: MIT Press.

Bijker, W., Hughes, T. and Pinch, T. (1987) *The Social Construction of Technological Systems: new directions in the sociology and history of technology*, Cambridge, MA: MIT Press.

Boczkowski, P. and Lievrouw, L.A. (2008) 'Bridging STS and communication studies: research on media and information technologies', in E. Hackett, O. Amsterdamska, M. Lynch and J. Wajcman (eds) *The Handbook of Science and Technologies Studies*, Cambridge, MA: MIT Press.

Bologna, C. (1992) *Flatus vocis: Metafisica e antropologia della voce*. Bologna: Il Mulino.

Castells, M., Fernandez-Ardevol, M., Qiu, J. and Sey, A. (2007) *Mobile Communication and Society: a global perspective*, Cambridge, MA: MIT Press.

Chu, W.C.R. (2008) 'The dynamics of cyber China: the characteristics of Chinese ICT use', *Knowledge, Technology & Policy*, 21(1): 29–35.

Claisse, G. (1988) 'Téléphone, communication et société: pratiques et critiques', *Communication*, 2: 97–118.

Claisse, G. and Rowe, F. (1993) 'Téléphone, communication et sociabilité: des pratiques résidentielles différenciées', *Sociétés contemporaines*, 2: 165–89.

Claisse, G., Vergnaud, T. and Rowe, F. (1985) *Téléphone, Communication et Société. Recherches sur l'utilisation domestique du téléphone*, Lyon: CNRS-ENTPE.

De Gournay, C. (2002) 'Pretence of intimacy in France', in J. Katz and M. Aakhus (eds) *Perpetual Contact: mobile communication, private talk, public performance*, Cambridge: Cambridge University Press.

De Sola Pool, I. (1977) *The Social Impact of the Telephone*, Cambridge, MA: MIT Press.

—— (1984) *Technologies of Freedom*, Cambridge, MA: Harvard University Press.

Donner, J. (2008) 'Research approaches to mobile use in the developing world: a review of the literature', *The Information Society*, 24(3): 140–59.

Dordick, H. and LaRose, R. (1992) *The Telephone in Daily Life: a study of personal telephone use*, The Bell Atlantic Corporation.

Fortunati, L. (1981) *L'arcano della riproduzione*, Venezia: Marsilio; trans. (1995) *The Arcane of Reproduction*, New York: Autonomedia.

—— (ed.) (1995) *Gli italiani al telefono*, Milano: Angeli.

—— (2005) 'The mobile phone as a technological artefact', in P. Glotz, S. Berschi and C. Locke (eds) *Thumb Culture: the meaning of mobile phones for society*, Bielefeld: Transcript Verlag.

—— (2006) 'User design and the democratization of the mobile phone', *First Monday* 11(special issue 7). Online. Available: http://firstmonday.org/htbin/cgiwrap/bin/ojs/index.php/fm/article/view/1615/1530 (accessed 12 January 2011).

Fortunati, L., Gebhardt, J., Vincent, J., Petrovčič, A. and Vershinskaya, O. (2010) *Interacting with Broadband Society*, Berlin: Peter Lang.

Gebhardt, J. (2010) 'Alfred Schütz and the media: the inter-subjective constitution of mediated interpersonal communication in everyday life', in L. Fortunati, J. Gebhardt, J. Vincent, A. Petrovčič and O. Vershinskaya (eds) *Interacting with Broadband Society*, Berlin: Peter Lang.

Gebhardt, J., Greif, H., Raycheva, L., Lobet-Maris, C. and Lasen, A. (eds) (2010) *Experiencing Broadband Society*, Berlin: Peter Lang.

Geser, H. (2004) *Towards a Sociological Theory of the Mobile Phone*. Online. Available: http://socio.ch/mobile/t_geser1.htm (accessed 15 January 2010).

Glotz, P., Bertschi, S. and Locke, C. (eds) (2005) *Thumb Culture: the meaning of mobile phones for society*, New Brunswick, NJ: Transaction.

Goggin, G. (2006) *Cell-Phone Culture: mobile technology in everyday life*, London: Routledge.

—— (ed.) (2008) *Mobile Phone Cultures*, London: Routledge.

Goggin, G. and Hjorth, L. (eds) (2009) *Mobile Technologies: from telecommunications to media*, New York: Routledge.

Haddon, L. (1994) 'The phone in the home: ambiguity, conflict and change', paper presented at the *COST 248 'The European Telecom User Meeting'*, Lund, Sweden, April.

—— (1998) 'Il controllo della comunicazione. Imposizione di limiti nell'uso del telefono', in L. Fortunati (ed.) *Telecomunicando in Europa*, Milano: Angeli.

—— (2004) *Information and Communication Technology in Everyday Life: a concise introduction and research guide*, New York: Berg.

Haddon, L., Mante, E., Sapio, B., Kommonen, K-H., Fortunati, L. and Kant, A. (eds) (2005) *Everyday Innovators: researching the role of users in shaping ICTs*, Dordrecht: Springer.

Hartmann, M., Rossler, P. and Höflich, J.R. (eds) (2008) *After the Mobile Phone: social changes and the development of mobile communication*, Berlin: Frank & Timme.

Hjorth, L. (2005) 'Postal presence: a study on mobile customisation and gender in Melbourne', in P. Glotz, S. Berschi and C. Locke (eds) *Thumb Culture: the meaning of mobile phones for society*, Bielefeld: Transcript Verlag.

Höflich, J. and Gebhardt, J. (eds) (2005) *Mobile Kommunikation. Perspektiven und Forschungsfelder*, Frankfurt am Main: Peter Lang.

Höflich, J. and Hartmann, M. (eds) (2006) *An Ethnographic View on Mobile Communication*, Berlin: Frank & Timme.

Katz, J. (1999) *Connections: social and cultural studies of the telephone in American life*, New Brunswick, NJ: Transaction Publishers.

—— (ed.) (2003) *Machines that Become Us*, New Brunswick, NJ: Transaction Press.

Katz, J. and Aakhus, M. (eds) (2002) *Perpetual Contact: mobile communication, private talk, public performance*, Cambridge: Cambridge University Press.

Kim, P. and Sawhney, H. (2002) 'A machine-like new medium: theoretical examination of interactive TV', *Media, Culture & Society*, 24(2): 217–33.

Lange, U. (1991) 'The Berlin telephone study: an overview', in A. Moyal (ed.) *Research on the Domestic Telephone*, Melbourne: Circit.

Lange, U., Beck, K. and Zerdick, A. (eds) (1989) and (1990) *Telefon und Gesellschaft*, Vols 1 & 2, Berlin: Volker Spiess.

Lasen, A. (2010) 'Mobile culture and subjectivities: an example of the shared agency between people and technology', in L. Fortunati, J. Gebhardt, J. Vincent, A. Petrovčič and O. Vershinskaya (eds) *Interacting with Broadband Society*, Berlin: Peter Lang.

Law, P.L. and Chu, W.C.R. (2008a) 'ICTs and China: an introduction', *Knowledge, Technology and Policy*, 21(1): 3–7.

—— (2008b) 'ICTs and migrant workers in contemporary China', *Knowledge, Technology and Policy*, 21(2): 43–45.

Law, P.L., Fortunati, L. and Yang, S. (eds) (2006) *New Technologies in Global Societies*, Hackensack, NJ: World Scientific.

Licoppe, C. and Smoreda, Z. (2005) 'Rhythms and ties: towards a pragmatics of technologically-mediated sociability', in R. Kraut, M. Brynin and S. Kiesler (eds) *Computers, Phones and the Internet: domesticating information technologies*, Oxford: Oxford University Press.

Ling, R. (2004) *The Mobile Connection: the cell phone's impact on society*, San Francisco: Morgan Kaufmann.

—— (2008) *New Tech, New Ties: how mobile communication is reshaping social cohesion*, Cambridge, MA: MIT Press.

Ling, R. and Pedersen, R.E. (eds) (2005) *Mobile Communication: re-negotiation of the social sphere*, London: Springer.

Manovich, L. (2001) *The Language of New Media*, Cambridge, MA: MIT Press.

Moyal, A. (1992) 'The gendered use of the telephone: an Australian case study', *Media, Culture and Society*, 14(1): 51–72.

Nyíri, K. (ed.) (2003) *Mobile Democracy: essays on society, self and politics*, Vienna: Passagen Verlag.

—— (ed.) (2005) *A Sense of Place: the global and the local in mobile communication*, Vienna: Passagen Verlag.

—— (ed.) (2008) *Integration and Ubiquity: towards a philosophy of telecommunications convergence*, Wien: Passagen Verlag.

Oudshoorn, N. and Pinch, T. (2003) *How Users Matter: the co-construction of users and technologies*, Cambridge, MA: MIT Press.

—— (2008) 'User–technology relationships: some recent developments', in E.J. Hackett, O. Amsterdamska, M. Lynch and J. Wajcman (eds) *The Handbook of Science and Technology Studies*, Cambridge, MA: MIT Press.

Pertierra, R. (ed.) (2007) *The Social Construction and Usage of Communication Technologies: Asian and European experiences*, Diliman, Quezon City: The University of the Philippines Press.

Pertierra, R., Ugarte, E.F., Pingol, A., Hernandez, J. and Dacanay, N.L. (2002) *Txt-ing Selves: cellphones and Philippine modernity*, Manila: De La Salle University Press.

Rakow, L. (1992) *Gender on the Line: women, the telephone and community life*, Urbana: University of Illinois Press.

Rakow, L. and Navarro, V. (1993) 'Remote mothering and the parallel shift: women meet the cellular telephone', *Critical Studies in Mass Communication*, 10(2): 144–57.

Rogers, E. (2003) *Diffusion of Innovations*, New York: Free Press.

Rutter, D. (1987) *Communicating by Telephone*, Oxford: Pergamon Press.

Salazar, P-J. (1987) 'Formes de la voix', *Sociétés*, (16):16.

Sawhney, H. (2007) 'Strategies for increasing the conceptual yield of new technologies research', *Communication Monographs*, 74(3): 395–401.

Schegloff, E. (2002) 'Beginnings in the telephone', in J. Katz and M. Aakhus (eds) *Perpetual Contact: mobile communication, private talk, public performance*, Cambridge: Cambridge University Press.

Schramm, W. (1977) *Big Media, Little Media: tools and technologies for instruction*, Beverly Hills, CA: Sage.

Schütz, A. and Luckmann, T. (2003) *Strukturen der Lebenswelt*, Frankfurt: Suhrkamp.

Silverstone, R. (2006) 'Domesticating domestication: reflections on the life of a concept', in T. Berker, M. Hartmann, Y. Punie and K.J. Ward (eds) *Domestication of Media and Technology*, Maidenhead: Open University Press.

Silverstone, R. and Haddon, L. (1996) 'Design and the domestication of information and communication technologies: technical change and everyday life', in R. Mansell and R. Silverstone (eds) *Communication by Design: the politics of information and communication technologies*, Oxford: Oxford University Press.
Silverstone, R. and Hirsch, E. (1992) *Consuming Technologies: media and information in domestic space*, London: Routledge.
Sørensen, C. (2006) *Mobile Phones at Work*, London: Carphone Warehouse.
Sørensen, C. and Pica, D. (2005) 'Tales from the police: rhythms of interaction with mobile technologies', *Information and Organization*, 15(2): 125–49.
Star, S. and Bowker, G. (2002) 'How to infrastructure', in L.A. Lievrouw and S. Livingstone (eds) *Handbook of New Media*, London: Sage.
Van Dijk, J. (1999) *The Network Society: an introduction to the social aspects of new media*, London: Sage.
Vincent, J. (2003) 'Emotions and mobile phones', in K. Nyíri (ed.) *Mobile Democracy: essays on society, self and politics*, Vienna: Passagen Verlag.
Vincent, J. and Fortunati, L. (eds) (2009) *Electronic Mediated Emotions: the mediation of emotion via information and communication technologies*, Oxford: Peter Lang.
Virilio, P. (1992) 'Big optics', in P. Weibel (ed.) *On Justifying the Hypothetical Nature of Art and the Nonidenticality within the Object World*, Köln: König.
Wajcman, J., Bittman, M. and Brown, J. (2009) 'Intimate connections: the impact of the mobile phone on work life boundaries', in G. Goggin and L. Hjorth (eds) *Mobile Technologies: from telecommunications to media*, New York: Routledge.
Ward, K. (2006) 'The bald guy just ate an orange: domestication, work and home', in T. Berker, Y. Punie, M. Hartmann and K. Ward (eds) *Domestication of Media and Technology*, Maidenhead: Open University Press.
Williams, F. (1982) *The Communications Revolution*, Beverly Hills, CA: Sage.
Wiredu, G. and Sørensen, C. (2006) 'The dynamics of control and mobile computing in distributed activities', *European Journal of Information Systems*, 15(3): 307–19.
Yung, V. (2005) 'The construction of symbolic values of the mobile phone in the Hong Kong Chinese print media', in R. Ling and P.E. Pedersen (eds) *Mobile Communications: re-negotiation of the social sphere*, London: Springer-Verlag.

3 Riding the double juggernaut

Depicting the Chinese mobile communication context

Rodney Wai-chi Chu

Introduction

To borrow from Katz and Aakhus (2002: 1), we can say the mobile phone, rather than the telephone in general as they claim, has become a fixture of everyday life in the twenty-first century. This at least is the case in Mainland China, as shown in Table 1.1 of this volume, where we can observe how land phone users are outnumbered by mobile phone users despite the fact that the former developed at a much earlier time. Precisely put, mobile telephony as an information and communication technology (ICT) device is a comparatively recent phenomenon. But whereas its development as a communication tool in the first world is more of a linear progress, the emergence and popularization of the device in China is a leapfrog because of the acute economic advancement of this communist regime. The Western sociological analyses have built up a handful of substantive studies concerning the cyberization of society, and of course there are mounting studies related to industrialization. However, a direct transplant of the literature to depict the Chinese scenario may call for some cautions, as China has been experiencing a process of compressed modernity which makes her experience in evolving into the modernization route somewhat different from that of the West. The nature of such compressed modernity is characterized by the rapid industrialization and cyberization of the place in almost a simultaneous manner. This process we name a 'double juggernaut ride', and the current discussion is to elaborate this claim.

We unfold the agenda by exploring first the meaning of reform to the Chinese leadership. This requires much deeper comprehension of the ideological concern of the ruling elites, where the notion of 'rule by morality' is having a cultural tradition instead of just merely an official rhetoric for political stability. Such an ideological touch helps us to come closer to terms with some ICT practices in China. In the next section, we move into an analysis of the interpretation of the West toward China's reform endeavour. This has laid down the foundation of why the 'double juggernaut ride' could be overlooked in the course of investigation. Eventually, when examining the related studies on Chinese soil in the communication field in general, and that of mobile phone studies in particular, we witness a preference of adopting a political sociology approach in the research orientation and foci in this area, which is not the case if we look back to the mobile phone study literature in the West. The discussion will end with an elaboration of a line of practice that has put the cultural realm back into attention, an approach which the author also adopts. We believe this emphasis is essential in examining mobile phone use in China in the midst of her reform and modernization process characterized by the 'double juggernaut ride'. We want to maintain that the *density* of experience (i.e. coming up with modern capitalism and cyberization at the same time) within a *compressed period* when the society has a thick *cultural heritage* of its own, and yet is still under a *dictating rule*, is complicated and sufficient enough to consolidate a kind of mobile phone use with Chinese characteristics, and such a thesis may contribute not only to a better grasp of mobile phone use in, but also the social fabric of, China in its current form.

Revisiting the meaning of reform to the Chinese leadership

To start with, in the eyes of the West, there is a general belief in a straightforward relationship between media reform and state democratization, as a book subtitle in this area of study shows – *Democratizing the Media, Democratizing the State* (Price *et al.* 2002). Thus, telecommunication and mass media as industries are more closely related to preservation of citizens' rights to be informed and to communicate freely, both of which are conditions essential to political democracy (Zhao 2000: 43). The Communist statesmen of course are not unfamiliar with such a line of thought, and this somewhat explains why media reform as a whole only launched almost after a decade of the opening of the country. The history of mobile phone development in China (Wang and Cheng in this volume, Chapter 5), which is one aspect of media reform, must be set against this context (Guan 2003; Lynch 2000). But more importantly, it is what sort of meaning the Communist leadership is giving to the notion of 'reform', and the *timing* of China's reform, that makes China so interesting.

What was in the minds of the post-Mao statesmen regarding the purpose of carrying out reform in China? And what are the consequences of that? For those in favour of a class-based critique, the common view is that Chinese reforms, focusing on capital marketization, have brought about consequences such as income polarization, increased poverty, and intensified exploitation (Pun 2005; Wu 2005). Eventually, economic reform without political reform would only lead to distorted development, and so some would say it is just wishful thinking of the Chinese statesmen to assume political authoritarianism is the price that the Chinese populace is willing to pay in exchange for economic development (Lee and Hsing 2009). The Chinese leadership obviously would interpret otherwise: economic liberalization may be a requirement of growth, but political liberalization is a threat to stability (Mueller and Tan 1996). Looking back at the literature of the previous cohort of a group of China watchers, it was argued in the mid-1980s that it would be fairer to refer to the reformist wing under Deng Xiaoping as the 'practice faction' instead of 'pragmatists': the reform aimed to improve rather than abolish Communist rule (Schram 1984: 2), and if there was any reform of the polity, it should be understood not for its ending, but its strengthening

(Bachman 1985: 163; Gittings 1989: 193). In light of these two polarized views of understanding, we have to go deeper into the Chinese tradition for a better grasp of the logic of statesmen.

Bakken argues there is a great misunderstanding of Chinese reforms as seen in the West. By describing the Chinese society as an 'exemplary society', he maintains that the Chinese type of modernization seems to follow a deeply rooted pattern of 'controlled change' aimed at taming and binding the path as well as the pace of the juggernaut toward modernity. In this process, tradition and modernity interlink in a system of social control where 'tradition' is seen as serving transforming purposes and 'modernization' can mean stability and order (Bakken 2000: 4–5). In this light, concepts like 'social engineering' (*shehui gongcheng*), 'moral construction' (*diode jianshe*), and 'two civilizations – material and spiritual' (*wezhi wenming* and *jingshen wenming*), which are so frequently mentioned in the official press and which are so popular among political or technocratic elites, are not empty slogans in themselves, but could rather be a striving for 'rule by morality' in the course of the capitalist ride (ibid.: 50–54).

Bakken's message has a timely meaning if we look further into what is meant by the modernization of technology to the Chinese leadership. Zhao has been providing a relatively detailed account linking up the ICT development in China with the broader concerns mentioned above (Zhao 1998, 2000, 2007, 2008; Zhao and Duffy 2007; Zhao and Schiller 2001). Simply put, Zhao maintains that China's military-led techno-nationalism has had a profound impact on her 'digital revolution' during the post-Mao era (Zhao 2007: 96). One of the pressing issues faced by the post-Mao leadership was to survive militarily in the age of high-tech wars and nuclear deterrence. An obvious result was the launching of a comprehensive hi-tech development programme known as the 863 Plan in March of 1986 (thus the name of the plan) in response to the Reagan administration's Star Wars initiative (ibid.: 97). The plan was the base for China's development of information technology. As such, development of ICTs is seen as instrumental to the state's modernization of its military and surveillance capabilities, and with this backdrop one may come to terms with a paradoxical feature of China's digital revolution: on the one hand, ICTs have been progressively promoted and widely, although unevenly, diffused among the population; on the other hand, state control over content and access of all kinds has never been far from the minds of the statesmen (ibid.: 92). To some extent, this echoes the view of media reform as undergoing a stop–go cycle in the 1980s (Chan and Qiu 2002: 33–34), as behind the pattern of reform and retrenchment lies the fear of the Chinese Communist Party leadership that freedom in related areas (from the conventional press freedom to freedom of expression in the cyber world) will pose a threat to their political power.

Such understanding may help us to witness the series of state acts that may show a continued logic: the 'anti-spiritual pollution campaign' lasting from October 1983 to February 1984, as well as the 'anti-liberalization campaign' from January to May 1987, together showed us how the official rhetoric behind these campaigns was always moral in tone. As Metzger contends, the roots of civil society in China are unlike those of the West which lie in the protection of individual freedom from state intervention; rather, the roots are in the realm of thoughts embedded in intellectual thinking – the duty of the intellectuals is to articulate the true needs of the people, and this is a good intention from those above (Metzger 1994, 1998). It is with this backdrop that Chu and Cheng use the Green Dam Youth Escort Online Filtering Software (which demonstrates the 'control change' of the state) and the subsequent '2009 Declaration of the Anonymous Netizens' (which has been celebrated by the West as an illustration of emerging civic power) to show how quite a handful of Chinese ICT users have been thinking otherwise in response to the provocative view of the Declaration (Chu and Cheng 2010: 30–36). It is thus essential to examine how Chinese ICT users are embedded in a particular socio-political environment *and* cultural tradition that should not have been downplayed as a factor for understanding the ICT phenomenon in China.

Similar to Bakken, Zhao points out that the Western view of the opening of China – that the opening of the communications market to global capitalism will undermine the Party's authoritarian control and facilitate the emergence of a liberal democratic order – can be an expectation instead of a picture of what is really happening there (Zhao 2008: 143–44). For sure, outsiders may query whether China can get the economic benefits of transnational/digital capitalism without paying the social price (Zhao and Schiller 2001), but Zhao wants to remind us that the development and value of ICTs have to be analysed in concrete political, economic, social and cultural contexts. But now, it is time to turn also to the Western interpretation of China's telecommunication reform, and its constraint in capturing the essence of the 'double juggernaut ride' as experienced by China.

The interpretation of the telecommunication reform and the double juggernaut ride

The opening up of China in the late 1970s marked a golden opportunity for transnational capital to prepare for landing on this under-developed market, and there was the eagerness to come to terms with how a Communist regime could modernize itself and integrate into the capitalist world. With such a pre-set agenda and attitude, literature in the telecommunications field published since the reform policy has put the emphasis on relating telecommunication markets with the social development of China (DeWoskin 2001; Guan 2003; Lee 1997; Lin *et al.* 2001; Lu and Wong 2003; Yan and Pitt 2002). As time elapses, analyses using a 'political sociology' viewpoint, which studies the dilemmas faced by the ruling regime in its efforts to maintain authoritarian control in light of the opening up of the communications market, have started gaining momentum (see, for instance, Chan and Qiu 2002; Chu 1997; Harwit 2004; Harwit and Clark 2001; Lu and Weber 2007; Lynch 2000; Mueller and Tan 1996; Wu 2009). However, such overwhelming concern has somewhat sidelined the unique process of compressed modernity that China has undergone – her entering into the global economy at a time when the global world also marked a march into the cyber era.

This is a double juggernaut because all Chinese living through the 1980s to date have faced the rapid capitalization and cyberization process, and people of all walks of life, whether they have been psychologically prepared or not, more or less have to experience a full-scale socio-cultural and techno-economic upheaval. The kind of 'compressed' feeling was particularly strong in the 1990s. In 1992 Deng Xiaoping made his famous southern tour of China between January and February, which was generally regarded as a method of reasserting his economic policy after the crackdown in June 1989. Whereas the tour covered the deepening of reform, among other aspects, in the telecommunication field, the same period also testified to how quickly humans have stepped into the global cyber era because of the historic development of the World Wide Web. Even in the United States, the pace of development of Internet and web culture has been astonishing (Chu and Cheng 2010: 25). While social scientists in the West have been busy depicting the origins, meanings and implications of the cyber age in an already industrialized/modernized setting in the last one and a half decades or so, China, with her determination to re-enter into the global world, chooses to live with a parallel path of rapid development, and history has yet to relate how this double juggernaut of modern capitalism and cyberization will be ridden.

This unique moment is characterized by a few interesting features that mark the different research paths of China and the West in the ICT field. First, unlike the articulation by Fortunati in the previous chapter, which she regards as an issue that has to be dealt with, the discontinuity of fixed phone and mobile phone research is not problematic in China, as there was little attention given to the former at the start. This is mainly because, before the 1990s, the limited state investment in landlines across the country (He 1997: 68–78; Yan and Pitt 2002: 13–19) had resulted in private lines being very costly and very time-consuming to connect (Yu and Tng 2003: 187), thus preventing the 'fixed phone experience' from being a 'normal experience in the contemporary era' for the Chinese people. Second, as has been said above, a general assumption of the consequence of media reform is democratization (Golding 1990; Rozumilowicz 2002), and as Zhao maintains (Zhao 2000, 2008), the Western liberal democratic wing believe in and desire such changes. Thus the concern has been more from a political sociology point of view, and so it is difficult to place mobile phone research in China in the 'main tenets in mobile phone studies' or to analyse the mobile phone 'as a technological artefact' as has been captured in the previous chapter by Fortunati. Simply put, it seems the research orientation on ICTs in China is not so much of 'a sociological enquiry of humans and technology' as an end, but more of using ICT research as a means to shed light on the 'socio-political dynamics within a communist regime in its path of modernization'. Third, if we look at the matter from within, the 'double juggernaut ride' that the Chinese have been experiencing has to be understood as a unique experience that even contemporary social scientists would have to learn how to get acquainted with. Here, it may be a bit provocative, but not without grounds, to say that a direct transplantation of Western sociological frameworks and concepts to depict the cyberization of China may raise our eyebrows. This is because we are not merely talking about cultural sensitivity under the umbrella phrase of 'indigenization of social science knowledge' when there are already plenty of discussions (see, for instance, Alatas

S.F. 1993, 2000, 2003; Alatas S.H. 1974; Atal 1974, 1981; Lee 2000), but also how to deal with the social dynamics of people living under a parallel evolution of two forces: on the one hand, a compressed process of industrialization, or even transnational capitalism; on the other hand, the cyberization of a place that, just 30 years ago, was a closed environment to the external world. Lastly, against the above backdrops, it is interesting to see that, in the last decade or so, far more effort has been poured into the study of Internet-related aspects of China than into the mobile phone (see, for instance, Chase and Mulvenon 2002; Damm 2007; Foster and Goodman 2000; Herold 2008; Hughes and Wacker 2003; Liu 2010; MacKinnon 2008; Mengin 2004; Qiu 2004; Tai 2006; G. Yang 2006, 2007, 2008, 2009; Yang and Calhoun 2007). Besides the dominance of the political sociology approach that sees that Internet studies can reflect more on the notion of freedom of expression, anti-oppression, openness and the like, it is of interest to make some other guesses.

One thing is the convenience of assessing Internet behaviour. For example, in reflecting on his methodology of doing online research, Guobin Yang mentions his close monitoring of several online communities and even runs a personal blog using a pseudonym in order to gain a personal understanding of the blog culture in China (Yang 2009: 22). In contrast, studies on the mobile phone appear to be more demanding, at least in terms of requiring more resources for data collection, if a researcher wants to make sense of this ICT device. After all, the mobile phone has been seen as a kind of communication tool rather than as an expressive tool when compared with a blog or a forum. But we must not overlook the fact that the mobile phone has the feature of mobility, and the emergence of software like Twitter which, as a free social networking and micro-blogging service that enables its users to send and read messages in an instant manner (thus Twitter is also known as the 'Short Message Service (SMS) of the Internet'), has helped to move the mobile phone to a new plateau of social meaning. What is more, we must acknowledge the fact that text messages (or SMS) play a particular role in mobile phone culture in China, which is much reflected by the volume of its use as seen in Table 3.1. By the end of 2005, the volume of SMS messages in China had already accounted for more than 50 per cent of global SMS messages (He 2008: 184), and thus the significance of SMS in the cyber world of China should be beyond doubt. By now, we arrive at a juncture point: mobile phone study is important, but it has been under-researched on Chinese soil. Yet, what have been the major endeavours in this area, and what can be its future orientation with all that has been examined hitherto? We are about to revisit what has been done and look forward to the possible prospect of this research agenda.

China mobile phone studies: looking backward and forward

If we exclude those studies with a policy-analysis tone or adopting a political sociology perspective in examining mobile communication development in China, it is safe to claim that the attempts by Caporael and Xie (2003) as well as

Table 3.1 Volume of SMS messages sent/received by mobile phone users in China (2004–2010)

Period (year end)	*SMS volume*[a]	*Total number of mobile phone users during the period*[b]	*Average SMS messages sent/received per mobile phone*
2004	2,177.6	33,482	650
2005	3,046.5	39,343	774
2006	4,296.7	46,108	932
2007	5,921.0	54,729	1,082
2008	6,996.7	64,123	1,091
2009	7,727.0	74,738	1,033
2010	8,317.0	85,900	968

Source: Ministry of Industry and Information Technology, The People's Republic of China, various years. There was no data on SMS usage prior to the year 2004.

Notes

a Unit: 100 million, rounding up to the nearest 1 decimal point.

b Unit: 10,000 users, rounding up to the nearest digit.

Yu and Tng (2003) are the first to explore the cultural realm of mobile phone studies on Chinese/China. By making a comparison between American and Chinese subjects, Caporael and Xie discover that the Chinese users are seldom aware of the public/private division (see also the discussion of their study by Peng and Chu in this volume, Chapter 13), and thus account for a very different behavioural pattern of mobile phone use. On the other hand, Yu and Tng take the communication culture of China as the spearhead of their study, and point out that economic liberalization and the new consumerism have initiated the separation of work and home and expand the realms of the personal and individual. Both studies should have opened up a fascinating account of the cultural side of China ICT use, yet a literature search shows that their endeavours have not been continued and expanded on. On the other hand, two studies (He 2008; Yu 2004) on SMS use by mobile phone users in China are among the very few that focus on substantive mobile phone usage, but both discussions have put the emphasis on analysing the struggle/emergence of civil society under a communist regime, and thus limit the context of exploring the 'Chineseness' of ICT use to the political realm.

It is against this backdrop that Jack Qiu appears as an active figure in mobile phone studies, reflecting his latest endeavour in a single-author book (Qiu 2009). Qiu himself and the team that he has been working with have proposed the 'information have-less' as a category in China to set against those 'information haves' and 'information have-nots' in the evolving network society (Cartier *et al.* 2005: 9–10). Qiu then identifies the 'have-less' as people with lower socio-economic status, such as the unemployed, retirees, students and migrants, who have begun to adopt ICTs on a massive scale. The emergence of this group is caused by the availability of low-end digital technologies (for example, Little Smart [*xiaolingtong*]) together with the blossoming of cybercafés (*wanba*) and SMS usage; indeed, these factors are critical to enhance labour mobility and form a 'translocal network', which may bring about extended *guanxi* connections among the information have-less. In the end, this makes the emergence of a working-class ICT possible (Qiu 2007), and the rise of a working-class network society (Qiu 2009: 2), or what is referred to as 'network labor' (ibid.: 5). This emphasis is important, as Qiu states that the answer to Manuel Castells' comment that 'China is constructing a different model of network society' (Castells 2008: 6 in Qiu 2009: 236) lies in this term.

Qiu's endeavour has aroused a few interesting concerns. First, by 2009, Qiu uses the phrase 'network labor' to refer to the 'information have-less' (i.e. the internal migrants and laid-off workers) which he puts at 177 million (Qiu 2009: 4). However, this figure accounts for less than 50 per cent of the total number of 'information have-less' (ibid.), so how should we weigh the significance of this group against this whole category? Second, there is a problem in understanding this sub-category of migrant workers itself. As a prominent figure in the study of migrant workers in China, Murphy points out that it is problematic to view them as economic labourers belonging to a homogeneous economic group or stratum (*dagonzu*) and thus overlook the fact that they are complex social beings – or heterogeneous individuals – who participate in various social relationships (Murphy 2009: 1). In an early study by Chu and Yang (2006), they maintain the cohort of migrant workers of 2000 (they term them 'new migrant workers') is different from those of the 1990s, but we must also pay attention to the socio-cultural backgrounds of this new group – they work in the city not merely to survive, but more for adventure (Chu and Yang 2006: 226–27). The aspiration for city life may be frustrating in terms of an unshaken rural identity, but the owning of a phone and the controlling of the technology when using it can counterbalance their low economic status and spartan living environment: among other things, owning a cell phone helps rebuild self-esteem (ibid.: 229), and there is no other consumer durable that can be held in the hand and used to express feelings, confirm a sense of freedom, and offer assurance of one's existence as a subject in a rootless time–space (ibid.: 230). This goes *beyond and behind* a 'class analysis' concerning who are the 'haves', 'have-nots' and 'have-less', and also captures a kind of life choice and life preference of a generation who are frustrated with the structural constraints they are facing, and yet choosing to live excitingly in a cyber mood as a means to put more emphasis on their individual freedom and enjoyment instead of how to strive for a living (Law and Peng 2008); as such, how far can we still use class analysis alone to depict the situation?

Low-end wireless technologies (such as Little Smart) are interpreted as a favourable condition for the have-less class in Qiu's eyes. Yet their cheap rates, as Chu and Yang contend, do not necessarily mean that the workers are spending less, as a cheap rate can easily generate more usage, as is seen in the bills and phone usage plans of their interviewees (ibid.: 228). A more interesting point is the fate of Little Smart. Even though Qiu points out at the end of his discussion that Little Smart faced a drastic decline in ARPU (average revenue per user) after 2004 (Qiu 2007: 916), this decline worsened in the following years: the total number of users fell to below 70 million by the end of 2008, and the whole service was terminated by 2011 (see Wang and Cheng in this volume, Chapter 5). In fact, according to the field data of Law and Peng of 2007, the growing number of odd-job workers demands a mobile service that gives them wider coverage and the required autonomy to be contacted instantly. Pagers are no longer useful for wildcat moves from one workplace to another. The mobile phone fulfils these new needs; the Little Smart suffers from limited mobility extension and its region-specific nature, whereas odd-jobbers may receive orders from different zones and regions (Law and Peng 2008: 61). The issue at stake here is that price may easily become a major concern if we are preoccupied by a class analysis; but if we open the space for the cultural dimension to step into the scene, the interpretation of the phenomenon can be more dynamic. For instance, one of the factors explaining the massive consumption and usage patterns of the mobile phone by migrant workers is a kind of 'social competition' that they carry out among themselves, but it is a pragmatic one: among people of similar background, the more advanced the model you have, the more *mianzi* (face) you gain. In a place that you do not and will never belong to, the owning or not of a mobile phone at the first level, and the owning of an up-to-date model instead of an obsolete one on a second level, is a symbol of basic achievement, a kind of social position that can help tackle the psychological emptiness of this group of rootless people (Chu and Yang 2006: 232). This is an entirely different dimension of seeing how culture and technology interface with one another: the mobile phone as an ICT device can extend their resource networks, can open up their cyber experience, can fill their psychological emptiness, and can create a social position. Such exposure is hardly the experience of the preceding cohort of migrant workers, because they were only riding the single juggernaut of industrialization in the mid-1980s to mid-1990s, far less sophisticated than the new migrant workers that have to face a double ride after entering the new millennium when ICT life in China sped up.

The emphasis on the cultural realm is not an attempt to belittle the class analysis; on the contrary, we hope to get a balanced view on the discourse of what is happening in China in her going-cyber process, which has obviously been overwhelmed by a political sociology orientation hitherto. After all, it is always tempting to examine the socio-political tension of a Communist state like China, not to mention the fact that she has become the world's spearhead for economic recovery since the worldwide economic crisis struck the globe in the summer of 2008. What is more, as mentioned above, telecommunication reforms are in themselves sensitive to ideas like openness, transparency and participation – all so much related to a soul-searching democratic urge. But if we hope to capture what is really happening instead of sticking with what we expect to happen, then we have to face squarely that the Chinese context cannot be captured only with a political sociology perspective. What exactly is influencing the normative basis of the people forces the cultural realm to the forefront.

Fairly stated, until recently, our group of researchers has increased the momentum of putting the cultural realm in a much more visible platform when examining ICT practices in China. Law and Chu in the series of articles recruited for the two special issues in 2008 (Law and Chu 2008a, 2008b) are obvious attempts of such. The two authors in one of the introductions bring forth the long-held attitudes in the social history of the Chinese toward science and technology so as to shed light on a better understanding of the spirit of ICT culture in China (Law and Chu 2008a). The Confucian philosophy sees that the purpose of pursuing knowledge of nature is for maintaining a harmonious relationship between man and nature instead of conquering it, and so there is no hunger for scientific/technological devices for life betterment (ibid.: 3–4). The almost total lack of science fiction in the literature of China is another piece of evidence showing the Chinese seldom consider machines and technology as important elements in their lives (ibid.: 5). From such a cultural heritage, and yet intermingled with such a speedy process of industrialization and cyberization, it would be difficult not to rethink how the subsequent startling influx of technological products and the rapid diffusion of ICTs into different modes of living could bring about a different cyber experience to the Chinese when comparing with their Western counterparts. The matter concerns the degree of difference.

So the study of Fortunati and others points out how the cultural differences between the West and China have affected their attitudes toward the mobile phone (Fortunati *et al.* 2008); Chu (2008) reveals how the Chinese concept of relational self shapes mobile communication among new migrant workers and young white-collars; Peng (2008) depicts how Internet use may help the new migrant workers to experience city life and satisfy their instrumental and expressive needs; K. Yang (2008) outlines the interesting relationship between the use of new technologies for the maintaining of traditional social relations; Cheng uses the call for 'boycott of Carrefour' spread through the public forum of the Internet and SMS messages via mobile phone to show the complication of viewing the Internet as promoting democracy in China (Cheng 2009); and on another occasion, Cheng shows the latest deep struggle the migrant workers have experienced in constructing social relations through mobile phone usage via the newly emerged Mobile QQ – a combination of mobile phones and computer instant messaging technology, thus depicting a new scenario of understanding the 'relational self' in Chinese culture (Cheng in press). In a nutshell, these quoted examples are attempts to capture some of the unique Chinese ICT behaviours, and the wider context of this is how people from different walks of life learn to live with a cyber era and transform some of their cultural practices into it, thus ending up with ICT behaviours sometimes compatible with their Western counterparts, and yet sometimes unique and deserving particular treatment. Needless to say, this edited volume is just another effort to continue this agenda.

Like what Fortunati concludes in the final paragraph of the preceding chapter, the previous pages of this chapter have also been sweeping. But the idea is to show how the double juggernaut ride is the essence of comprehending the kind of normative basis the contemporary Chinese are now situated in. China enters into the global economy at a time when the global world also marks a march into the cyber era, and within herself, she has her own political, cultural and economic forces to struggle with. People of Mainland China living through the 1980s to now are prisoners of this condensed mood, and witnesses of this juggernaut ride. The thesis reflects a complicated picture consisting of how the East meets the

West, how humans interplay with technology, and how tradition coincides with modernity. Mobile phone studies on China/Chinese as a research area are at the forefront to capture and demonstrate the dynamics the people are living with because the mobile phone as an ICT device has become, due to technological advancement, more and more comprehensive in drawing people, events and values together. It will be a convenient platform to bridge with various factors, but it will also be a difficult stage to ride, in the forever evolving and advancing cyber world – particularly in a leapfrogging China.

References

Alatas, S.F. (1993) 'On the indigenization of academic discourse', *Alternatives*, 18: 307–38.
—— (2000) 'An introduction to the idea of alternative discourse', *Southeast Asian Journal of Social Science*, 28(1): 1–12.
—— (2003) 'Academic dependency and the global division of labour in the social sciences', *Current Sociology*, 51(6): 599–613.
Alatas, S.H. (1974) 'The captive mind and creative development', *International Social Science Journal*, 26(4): 691–700.
Atal, Y. (ed.) (1974) *Social Sciences in Asia*, New Delhi: Abhinav Publications.
—— (1981) 'The call for indigenization', *International Social Science Journal*, 33(1): 189–97.
Bachman, D.M. (1985) *Chen Yun and the Chinese Political System*, Berkeley: Institute of East Asian Studies, University of California.
Bakken, B. (2000) *The Exemplary Society: human improvement, social control and the dangers of modernity in China*, Oxford: Oxford University Press.
Caporael, L.R. and Xie, B. (2003) 'Breaking time and place: mobile technologies and reconstituted identities', in J.E. Katz (ed.) *Machines that Become Us: the social context of personal communication technology*, New Brunswick, NJ: Transaction Publishers.
Cartier, C., Castells, M. and Qiu, J. (2005) 'The information have-less: inequality, mobility, and translocal networks in Chinese cities', *Studies in Comparative International Development*, 40(2): 9–34.
Castells, M. (2008) 'Interview with Manuel Castells', *Chinese Journal of Communication*, 1(1): 3–6.
Chan, J.M. and Qiu, J. (2002) 'China: media liberalization under authoritarianism', in M.E. Price, B. Rozumilowicz and S.G. Verhulst (eds) *Media Reform: democratizing the media, democratizing the state*, London: Routledge.
Chase, M.S. and Mulvenon, J.C. (2002) *You've Got Dissent! Chinese dissident use of the Internet and Beijing's counter strategies*, Santa Monica, CA: RAND.
Cheng, C.T. (2009) 'New media and event: a case study on the power of the Internet', *Knowledge, Technology and Policy*, 22(1): 145–53.
—— (in press) 'Floating workers and mobile QQ: the struggle in the search for roots', in L. Fortunati, R. Pertierra and J. Vincent (eds) *Migrations, Diaspora, and Information Technology in Global Societies*, London: Routledge.
Chu, L.L. (1997) 'The political economy of the communication system in China', in P.S.N. Lee (ed.) *Telecommunications and Development in China*, Cresskill, NJ: Hampton Press.
Chu, W.C.R. (2008) 'The dynamics of cyber China: the characteristics of Chinese ICT use', *Knowledge, Technology and Policy*, 21(1): 29–35.
Chu, W.C.R. and Cheng, C.T. (2010) 'Cultural convulsions: examining the Chineseness of cyber China', in D.K. Herold and P. Marolt (eds) *Online Society in China: creating, celebrating, and instrumentalising the online carnival*, London: Routledge.

Chu, W.C.R. and Yang, S. (2006) 'Mobile phones and new migrant workers in a south China village: an initial analysis of the interplay between the "social" and "technological"', in P.L. Law, L. Fortunati and S. Yang (eds) *New Technologies in Global Societies*, Hackensack, NJ: World Scientific.

Damm, J. (2007) 'The Internet and the fragmentation of Chinese society', *Critical Asian Studies*, 39(2): 273–94.

DeWoskin, K. (2001) 'The WTO and telecommunications sector in China', *China Quarterly*, 167: 630–54.

Fortunati, L., Manganelli, A.M., Law, P. and Yang, S. (2008) 'Beijing calling ... modernisation and the social effects of new media in China', *Knowledge, Technology, and Policy*, 21(1): 19–27.

Foster, W. and Goodman, S.E. (2000) *The Diffusion of the Internet in China*, Stanford, CA: Stanford University, Center for International Security and Cooperation (CISAC).

Gittings, J. (1989) *China Changes Face: the road from revolution 1949–1989*, Oxford: Oxford University Press.

Golding, P. (1990) 'Political communication and citizenship: the media and democracy in an in-egalitarian social order', in M. Ferguson (ed.) *Public Communication: the new imperatives*, London: Sage.

Guan, S.Y. (2003) *China's Telecommunications Reforms: from monopoly towards competition*, New York: Nova.

Harwit, E. (2004) 'Spreading telecommunications to developing areas in China: telephones, the Internet and the digital divide', *China Quarterly*, 4: 1010–30.

Harwit, E. and Clark, D. (2001) 'Shaping the Internet in China: evolution of political control over network infrastructure and content', *Asian Survey*, 41(3): 377–408.

He, Z. (1997) 'A history of telecommunications in China: development and policy implications', in P.N.S. Lee (ed.) *Telecommunications and Development in China*, Cresskill, NJ: Hampton Press.

—— (2008) 'SMS in China: a major carrier of the nonofficial discourse universe', *The Information Society*, 24(3): 182–90.

Herold, D. (2008) 'Development of a civic society online? Internet vigilantism and state control in Chinese cyberspace', *Asian Journal of Global Societies*, 2(1): 26–37.

Hughes, C.R. and Wacker, G. (eds) (2003) *China and the Internet: politics of the digital leap forward*, London: RoutledgeCurzon.

Katz, J.E. and Aakhus, M. (2002) 'Introduction: framing the issues', in J.E. Katz and M. Aakhus (eds) *Perpetual Contact: mobile communication, private talk, public performance*, Cambridge: Cambridge University Press.

Law, P.L. and Chu, W.C.R. (2008a) 'ICTs and China: an introduction', *Knowledge, Technology and Policy*, 21(1): 3–7.

—— (2008b) 'ICTs and migrant workers in contemporary China', *Knowledge, Technology and Policy*, 21(2): 43–45.

Law, P.L. and Peng, Y. (2008) 'Mobile networks: migrant workers in Southern China', in J.E. Katz (ed.) *Handbook of Mobile Communication Studies*, Cambridge, MA: MIT Press.

Lee, C.K. and Hsing, Y.T. (2009) 'Social activism in China: agency and possibility', in C.K. Lee and Y.T. Hsing (eds) *Reclaiming Chinese Society: the new social activism*, London: Routledge.

Lee, P.S.N. (ed.) (1997) *Telecommunications and Development in China*, Cresskill, NJ: Hampton Press.

Lee, S.H. (2000) 'The rise of East Asia and East Asian social sciences' quest for self-identity', *Journal of World-Systems Research*, 6(3): 768–83.

Lin, J., Liang, X. and Wan, Y. (eds) (2001) *Telecommunications in China: development and prospects*, New York: Nova.

Liu, F. (2010) 'The norm of the "good" netizen and the construction of the "proper" wired self: the case of Chinese urban youth', *New Media & Society*, 13(1): 7–22.

Lu, D. and Wong, C.K. (2003) *China's Telecommunications Market: entering a new competitive age*, Cheltenham: Edward Elgar.

Lu, J. and Weber, I. (2007) 'State, power and mobile communication: a case study of China', *New Media & Society*, 9(6): 925–44.

Lynch, D.C. (2000) 'The nature and consequences of China's unique pattern of telecommunications development', in C.K. Lee (ed.) *Power, Money and Media: communication patterns and bureaucratic control in cultural China*, Evanston, IL: Northwestern University Press.

MacKinnon, R. (2008) 'Flatter world and thicker walls? Blogs, censorship, and civic discourse in China', *Public Choice*, 134(1–2): 47–65.

Mengin, F. (ed.) (2004) *Cyber China: reshaping national identities in the age of information*, New York: Palgrave Macmillan.

Metzger, T.A. (1994) 'Modern Chinese utopianism and the Western concept of civil society', *Asian Perspective*, 18(1): 1–6.

—— (1998) *The Western Concept of a Civil Society in the Context of Chinese History*, Stanford, CA: Hoover Institution on War, Revolution and Peace.

Mueller, M. and Tan, Z. (1996) *China in the Information Age: telecommunications and the dilemmas of reform*, Westport, CT: Praeger.

Murphy, R. (2009) 'Introduction: labor migration and social development in China', in R. Murphy (ed.) *Labour Migration and Social Development in Contemporary China*, London: Routledge.

Peng, Y. (2008) 'Internet use of migrant workers in the Pearl River Delta', *Knowledge, Technology and Policy*, 21(2): 47–54.

Price, M.E., Rozumilowicz, B. and Verhulst, S.G. (eds) (2002) *Media Reform: democratizing the media, democratizing the state*, London: Routledge.

Pun, N. (2005) *Made in China: women factory workers in a global workplace*, Durham, NC: Duke University Press.

Qiu, J.L. (2004) 'The Internet in China: technologies of freedom in a statist society', in M. Castells (ed.) *The Network Society: a cross-cultural perspective*, Cheltenham: Edward Elgar.

—— (2007) 'The accidental accomplishment of Little Smart: understanding the emergence of a working-class ICT', *New Media & Society*, 9(6): 903–23.

—— (2009) *Working-class Network Society: communication technology and the information have-less in urban China*, Cambridge, MA: The MIT Press.

Rozumilowicz, B. (2002) 'Democratic change: a theoretical perspective', in M.E. Price, B. Rozumilowicz and S.G. Verhulst (eds) *Media Reform: democratizing the media, democratizing the state*, London: Routledge.

Schram, S.R. (1984) 'Economics in command? Ideology and policy since the 3rd Plenum, 1978–1984', *China Quarterly*, 99: 417–61.

Tai, Z. (2006) *The Internet in China: cyberspace and civil society*, New York: Routledge.

Wu, I.S. (2009) *From the Fist to Invisible Hand: the uneven path of telecommunications reform in China*, Stanford, CT: Stanford University Press.

Wu, Y. (2005) 'Rethinking "capitalist restoration" in China', *Monthly Review: An Independent Socialist Magazine*, 57(6): 44–63.

Yan, X. and Pitt, D. (2002) *Chinese Telecommunications Policy*, Boston: Artech House.

Yang, G. (2006) 'The Internet and civil society in China: co-evolutionary dynamics and digital formations', in L. Dittmer and G. Liu (eds) *China's Deep Reform: domestic politics in transition*, Lanham, MD: Rowman & Littlefield.

—— (2007) 'How do Chinese civic associations respond to the Internet? Findings from a survey', *China Quarterly*, 189: 122–43.

—— (2008) 'Contention in cyberspace', in K. O'Brien (ed.) *Popular Protest in China*, Cambridge, MA: Harvard University Press.

—— (2009) *The Power of the Internet in China: citizen activism online*, New York: Columbia University Press.

Yang, G. and Calhoun, C. (2007) 'Media, civil society, and the rise of a green public sphere in China', *China Information*, 21(2): 211–36.

Yang, K. (2008) 'A preliminary study on the use of mobile phones amongst migrant workers in Beijing', *Knowledge, Technology and Policy*, 21(2): 65–72.

Yu, H. (2004) 'The power of thumbs: the politics of SMS in urban China', *Graduate Journal of Asia-Pacific Studies*, 2(2): 30–43.

Yu, L. and Tng, T.H. (2003) 'Culture and design for mobile phones for China', in J.E. Katz (ed.) *Machines that Become Us*, New Brunswick, NJ: Transaction Publishers.

Zhao, Y. (1998) *Media, Market, and Democracy in China: between the party line and the bottom line*, Champaign: University of Illinois Press.

—— (2000) 'Caught in the web: the public interest and the battle for control of China's information superhighway', *Info*, 2(1): 41–66.

—— (2007) 'After mobile phones, what? Re-embedding the social in China's digital revolution', *International Journal of Communication*, 1: 92–120.

—— (2008) *Communication in China: political economy, power and conflict*, Lanham, MD: Rowman & Littlefield.

Zhao, Y. and Duffy, R. (2007) 'Short-circuited? The communication labor struggles in China', in C. McKercher and V. Mosco (eds) *Knowledge Workers in the Information Society*, Lanham, MD: Lexington Books.

Zhao, Y. and Schiller, D. (2001) 'Dances with wolves? China's integration into digital capitalism', *Info*, 3(2): 137–51.

Part II

Theory, history and comparison

4 Mobile communication and the fourth communicative revolution[1]

Leopoldina Fortunati

Introduction

The third communicative and cultural revolution, that of the electronic media, is unfolding in a framework that has seen a shift from modernity to post-modernity which is further characterized by globalization (Myerson 2001). Modern or Fordist society has been subject to tumultuous changes (Appadurai 1990; Harvey 1989; Jameson 1991; Lyotard 1979; Spivak 1999; Tomlinson 1999). We quote just some of these: the end of great narrations such as control of the collective imagination, the fragmentation of genres and tastes, the enormous diversification of the supply of and demand for products, the demise of the factory as the dominant social model, the multiplication of work typologies, the proliferation of family models, the overturning of the relations between material and immaterial work in favour of the latter, the spread of the logic of reproduction in the production of exchange values, the global extension and integration of industrial and financial activities, and so forth.

The extended family of electronic media has played an important role in this shift from modernity to post-modernity, supporting, mediating and enhancing the shift. We can speak in fact of the third cultural and communicative revolution carried out by electronic media because through them deep social, economic and cultural changes have been mediated. But is there any one device that because of its communicative characteristics and potentialities plays a specific role in post-modernity? My thesis is that there is indeed one such device: the mobile phone. The mobile phone has a competitor in the computer, especially after the advent of the Internet. But among the extended family of electronic media we must distinguish the mobile phone from other media for the following reasons:

1 It is the technology that has had the fastest and widest diffusion globally, impacting on the social definition of 'space/time category and globalization'.
2 It is the technology that has opened the door to the personalization of the information and communication technologies (ICTs), enhancing the 'individualization' process.
3 It is the technology that has reduced the distance from, and thus enhanced, the 'technological hybridization of the human body'.
4 It is on the mobile phone that the most significant frontiers of 'technological, communicative and cultural convergence' are challenged.

It is evident that the mobile phone has been used to redefine fundamental processes that are at the core of post-modernity and globalization (Castells *et al.* 2007). These social processes were opened up or supported by means of the mobile phone. This means that people have seen in the mobile phone the opportunity to develop processes which were lacking or which could be exploited. Probably these premises already existed at the social and economic level to overcome modernity. What was lacking was the means to produce this shift at least in the four directions cited above.

The thesis, I repeat, that I wish to discuss in this chapter is that the mobile phone, for the reasons mentioned, has played a crucial role in the opening of this new stage in the history of culture and communication, which corresponds to the shift to post-modernity and globalization. This shift should be seen not as a stage in an evolutionary process but rather as a radical shift, one which we could call 'the fourth communicative revolution'. Cultural sociologists (Griswold 1994) usually distinguish three cultural and communicative revolutions: the introduction of chirographic writing, the development of the press, and finally the advent of electronic media, both bidirectional (telegraph, telephone, fax, computer, mobile phone) and unidirectional (radio, television, video-recorder, audio- and video-cassette, compact disc). Electronic media have taken us from the modern to the post-modern era.

However, this distinction between the three cultural and communicative revolutions, albeit very useful, risks being too generic. Beniger (1986) and Williams (1982) proposed further distinguishing between two different concentrations of innovations in the third revolution. The first would have taken place around the turn of the twentieth century, while the second would be still ongoing. Van Dijk (1999) urges caution when speaking of technological revolutions, as the history of technology is more an evolutionary than a revolutionary process. But the point is not what is new from a technological point of view in a specific revolution. The point is how and to what extent societies by means of a communicative technology change the structure of their social organizations, the quantity and quality of their social relationships, and the everyday life of individuals. The advent of the mobile represented a leap forward, putting in motion social phenomena that seem to justify reference to a fourth revolution in human communications. This fourth revolution accompanies the social and economic changes associated with post-modernity and globalization. In this chapter, my discourse will be based on a series of research projects carried out during this decade on the different aspects of the social role of the mobile phone. Each of the above mentioned topics – space/time categories, individualization, hybridization and convergence – will be presented in a section of its own.

Speed and spread of the diffusion of the mobile phone

Mobile phone technology has spread faster and more widely across the globe than other technologies, eclipsing even the Internet. In the old industrialized countries, no other technology has approached its speed or spread. The fixed telephone, as

Donner recalls, needed a century to reach the first billion subscribers, while the mobile phone achieved the same in only two decades (Donner 2008). Aronson argued that 'the differences between a society that has an effective telephone system and one that does not are as great as those between literate and preliterate societies' (Aronson 1977: 36). But this difference is even more apparent with the mobile phone, as it is the mobile that shows more radically this difference between modernity and post-modernity. Also, in the new industrialized and developing countries, no other technology has spread so far in so short a time. In China, for instance, there are more than half a billion mobile phones and in India around 350 million. This capacity of the mobile phone to spread in advanced countries as well as in Asia or in developing countries shows that the mobile phone is useful and convenient even in societies at different stages of industrialization.

The global diffusion of the mobile phone has made the strongest contribution to the social redefinition of the categories of space/time (Fortunati and Law in this volume, Chapter 7), or rather the annihilation of space by means of time, to use Marx's words. Given that space apparently is no longer such a problem, it is on time that all the attention is focused. Humankind, by means of the mobile phone, has experienced several temporal dimensions, such as that of presence/absence (Gergen 2002) and the concept of rhythm (Haddon 2004; Licoppe and Smoreda 2005; Ling 2004; Sørensen and Pica 2005), that are linked to the temporal determinations of interpersonal relationships (Werner and Haggard 1985).

Furthermore, the mobile phone has been appropriated globally for the following reasons. First, its orality enables it to cross the barrier of illiteracy. Second, its low costs, in comparison to the fixed line and the Internet, explain the very high volume of communicative exchanges that the mobile phone mediates. Third, its great flexibility means that it can be reinvented by different users and used in many ways and for various purposes. Lastly, the speed of the appropriation of the mobile phone means that it is highly valued by users, probably for the margin of freedom that it offers and also for the level of democratization that passes through this device (Nyiri 2003). The mobile has represented in fact the occasion for developing the process of democratization of interpersonal communicative access in society in spatial and temporal terms. For women and adolescents and ever more children (Oksman and Rautiainen 2003) this instrument has represented the opportunity of claiming their own communicative identity to make them traceable without going through the head of the family or being under family control. It has furthermore made possible the organization of political demonstrations which have contributed in many countries to redefining the political order (Castells *et al.* 2004; Rheingold 2002).

Unlike the computer/Internet, the diffusion of which registers great limitations in old industrialized countries and yet greater limitations in the new industrialized and developing countries, the mobile phone sees a large adhesion from users and potential users all over the world. It is the mobile phone that embodies the global zeitgeist (Katz and Aakhus 2002) to the extent that it might be considered as the emblem of globalization (Castells *et al.* 2007). This is so true that, although the mobile phone follows in the tradition of the fixed telephone of mediating local communications and contacts, it is perceived more and more as global. The meaning of this strategic role played by the mobile phone can be found in Marx's *Grundrisse*, where Marx speaks of the means of communication and transport as 'spheres of valorization' (Marx 1857–58/1968: 160ff., vol. II), but also in Gebhardt (2010) who, revisiting Alfred Schütz (1967), speaks of communication technologies as communicative working tools. In effect, all the ICTs – mobiles included – act at the core of globalization because they produce value.

Fortunati further developed this analysis in terms of the ICTs, and in particular the mobile phone, as work tools for reproduction and strategic tools of social labour (Fortunati 2007). From a technological point of view, the mobile phone is an intellective machine (Maldonado 1997) which forms part of the broader process of mechanization, virtualization and digitalization of the civil sphere and civil society. People have used the mobile phone to link the world of work better with that of the family, by rationalizing organization to their advantage and thus saving time, money and effort. But if we see it as a work tool, we will discover that its widespread use has taken on other dimensions. In fact it has had the effect of making people in offices and factories and even in the domestic sphere more productive, penetrating through the very pores of the working day of men, women, youth and children, sweeping away much rigidity and inertia, and eliminating many shadow areas in which people disappeared to 'take a breather' as it were, thus avoiding the continuity of command and control by the organization of work or family and, in general, social networks. The use of the mobile phone has had the unexpected effect of depriving workers and people in general both of the time and space of social disconnection and the thousand defence strategies which counted on this disconnection.

Individualization

The second reason for which the mobile phone might be considered the technology of the fourth communicative and cultural revolution is that it was the first ICT to enhance individualization. This role is particularly important because individualization is the key social process in modernity and post-modernity (Fortunati and Law in this volume, Chapter 7). In the individualization process and in the ever-greater development of individuality, the mobile phone has signalled an important shift. Many examples can be given: from adolescents who try and evade parental control to the management of illicit love affairs; from the elderly who feel more secure in walking alone around town, to travellers who have transformed the mobile phone into an anti-anxiety instrument. The mobile phone has changed since the first stage of its diffusion from being a mobile technology into a personal technology, thus beginning another story in the practices of use of technologies. Personal ownership and use of the mobile phone has resulted in its control. The control of the machine has led to more control of the sphere of interpersonal relationships mediated by the mobile phone. This has augmented the communication power of the individual. The identity of the other electronic media has always been more of the family or at least more collective. The path of the Internet, for instance, is very different from that of the mobile: the use of the computer is also generally an individual one, but often the household computer is used by more than one member of the family. We could say that, if the television has served to divide a family from others, the mobile phone has served, because of its personal use, to divide an individual from others, even within the same family. The mobile's being on the body has obviously accentuated its becoming individual.

In its turn, personal use of the mobile phone has supported the process of individualization, since the mobile is the technology that allowed individuals to face the new rigidity and social segregation that have been created following the destruction of social practices in the public space, which have been put into motion in response to the social movements of the 1960s and 1970s. The randomness of social relations that has followed has meant the implementation of a specific regimentation of what until recently was fluid and spontaneous. At the same time, women, children and adolescents have caught the ball on the rebound and won a fuller communicative identity. However, all – including men – have begun to use the mobile to develop social communication as little subjected to micro-controls as possible.

In Western societies individualization has historically meant reinforcement of individuals' personal will and freedom, to the detriment of their capacity for mediation with others (see for instance the rise in the number of divorces), the capacity for socialization (the post-modern individual spends more periods alone during his/her life) and the capacity for co-present sociality. Reinforcing one's ego means necessarily weakening that part of social relations that implies intimacy, dependence on others and subjection to micro-controls on the part of the circles of family members, relatives, friends, acquaintances, etc. In this context, the development of communication that has occurred with the spread of the mobile seems to have had the effect of removing obstacles and communicative difficulties between young and old, appeasing social tensions and conflicts which have been drowned in a river of words of little communicative weight. Maldonado recalls instead that chatting is nothing more than a way of hiding the lack of communication (Maldonado 2005: 31). He underlines how much more important it is to try and understand the historical, social and cultural realities and the distorting forces that lie behind this great production of words. Mobile communication may be seen as a significant political problem for the new world order because its explosion can diminish the power of individual communication and threatens to invalidate the meaning of individual freedom.

The mobile phone has contributed also to the blurring of the frontiers between the public and the private, in the end supporting the fall of the private man and woman. Take for example the use of the mobile in public space by 'cell-shouters'. The fact that people shout out private conversations when among an anonymous crowd agrees with and reinforces that great social process established in industrialized countries at the end of the 1960s with the student movements, workers and above all feminists, which consisted of a redefinition of what is public and what is private. The feminist slogan 'private is public' is a testimony to the overturning of the way the subject was seen then. That is, the distinction of and separation between the public sphere and private sphere was contested then through the redefinition of the private sphere as public sphere. As a consequence, these two spheres began to merge. Maldonado writes that the new world about to be born will not be that prophesied by R. Sennett (1977) of the 'decline of public man', but rather that of the 'decline of private man' (Maldonado 2005: 43), as already announced by Habermas in the premise of his *The Structural Transformation of the Public Sphere* (Habermas 1989). What is happening, in his opinion, is that the public individual is colonizing the private individual – that is, the private is increasingly the public domain (Maldonado 2005: 44). The fact is that information and communication technologies have allowed the de-localization both of the private (outdoor) and the public (indoor), which implies that private and public 'are now everywhere' (ibid.). This has had at least two consequences. On one hand, on the private front there has been the de-intimatization of interiority with the consequent emptying out of the intimate sphere (Habermas 1989). On the other, on the public front (intended not only as the state but also as the sphere of civil society) a new system has opened out. The ideological machines of the state, adds Maldonado (2005), have been joined by the extra-state and extraterritorial machines of global governance, which with sophisticated technologies of control can seriously threaten our privacy, with the effect of putting our social identity into more serious crisis.

The mobile was the first warning, together with fashion, of the rejection of privacy at the mass level. That is, from a certain point onwards there was a great impatience with privacy, which was perceived as isolation, segregation and exclusion from the public sphere. The social design that extended the bourgeois experience of domestic interiority to the masses was contested widely. In the meantime people had started to discover the body, to contest the privacy of nudity as segregation of knowledge in respect to the 'other'. The movements of the late 1960s tried to propose nudity as a normal social pact. Fashion, attentive to the new signals and social behaviour, caught on to this exhibition of nudity by proposing the nude look, deeper cleavages, and more recently the flaunting of underwear, such as bras and panties, thus breaking the order of over/under, and the like. Among the various technologies, the cinema and television took on this rejection of the private, showing nude bodies within narrative contexts. The computer extended this show of nude bodies at the purely informational and commercial level (Liu 2005).

The mobile has caused a further leap in this destruction of privacy, which is perceived now as isolation from others, segregation, diminished solidarity, and the like. By means of this instrument people have begun to speak of everything, including their most intimate thoughts, exhibiting them in the public space without any embarrassment. Washing dirty linen in public and publicizing private affairs have meant a strong social desire to share experiences of life and points of view, discourses which, by pure chance, happen also to be the basis and premise of political action.

Elimination of distance from the human body

The third reason that justifies the assessment of the mobile phone as the technology of the fourth and cultural revolution is the fact that the mobile is the device that first broke down the distance from the human body, opening a new ontology of the human being. The mobile phone, eliminating the traditional distance existing (even if in a variable measure) between ICTs and the human body, made in fact a new centrality of the body emerge in the paths of technological innovation. From this aspect is derived a certain hybridization between the technology and the human body (Fortunati *et al.* 2003; Katz 2003; Wajcman 2004). The closing in of technologies to the body and the hybridization between the human body and technologies are in any case a problem to further investigate and debate. The consequent fusion between the human body and technology redescribes the social relation that individuals have with their body and the inalienable possession of it, but also with their body as their first material, communicative and aesthetic support.

Paradoxically it was art, the queen of artifice, that was the first to realize the drawing near of the mobile to the body, and even the penetration of the body by technologies (subcutaneous microchips, etc.). Fashion afterwards threw open the door to wearable technologies. We may trace the beginnings of this process to the appearance of body art which shifted the gaze towards the surface of the body itself. If the creative mind that belongs to and is inscribed in a body begins to see itself as something observable, workable and expressible, then the process of alienation and objectivization of that body becomes unstoppable. The appropriation of the body as an artistic space became very soon appropriation of the body as an expressive and technological space, on a mass level. The body, from being an inexhaustible place of experience, is becoming an unknown land, a place of technical innovation, in which the concept of inside and outside is no longer clear-cut and in which nature and artificiality are even more mixed up. It is becoming a place of presence/absence and the place from which, by means of the mobile, we are constructing a mediated communicative relation with the world, always more global, at least at the level of perception and social representation.

But not only was art the first to realize the penetration of technologies inside the body; fashion, which is after all closely connected to art, also seems to have had an intuition of and adumbrated this process. As at the technological level the shift is being prepared from the wearable computer to the body as a definitive wearable computer (Gershenfeld 1999), as at the fashion level the communicative work of the fashion message has begun to be conducted on the body. By means of tattoos, piercing, wounding, and the like, communication has tended to become stable, while through vestimentary style it changes with the seasons. It is the body in its individual interpretation that has become now more or less fashionable.

However, if it is true that identity is constructed starting from the material body, the mobile has played an increasing role in the construction of identity and especially in the social presentation of the self (Ling and Pedersen 2005). For instance, if the naked body is an undifferentiated body and the social body is instead a differentiated and clothed body, which can give much information, the body that wears a mobile on it is a body that more strongly articulates its distance from nature and which has a more precise collocation in a social and communicative context. Now it is well known that the more bodies speak, the more they facilitate social organization and are therefore reassuring. The body supplied with a mobile is especially reassuring in respect to the generic clothed body because it offers a large amount of direct and indirect information as to who is using it. It is also a body that reassures, as when needed the mobile can be used as a security blanket (Wearing 2002). Finally, if the information that speaking bodies transmit follows precise norms and ritualizations and follows socially shared obligations, these bodies become even more reassuring, because they allow members of society greater mutual control.

Only health technologies and, among ICTs, only the mobile have eliminated the distance from the body, bringing about important psychological and social modifications in the relationship with our body and in the construction of identity and relations with the Other. The abolition of distance from the human body that the mobile has brought with it has affected the concept of identity, which is constructed starting from the capacity to trace precise boundaries between oneself and others. 'Our identity is conquered by means of our opening out towards others, but also through our closure before the intrusiveness of others' (Maldonado 2005: 36). This consideration must be extended also to technology, in that identity has been constructed historically also on the distinction of a distance of technology from the body, as has always been true for electronic media. This means that mobile identity is a more open and social identity, with perhaps some problems in building the distance from others.

The role of the mobile in the construction of identity, however, must not be over-estimated, because it is also true that the development of cooperation as a natural force of social labour in globalization actually requires the tribute of an ever more undifferentiated labour force, so that individuals are forced to under-develop their singularity. Baudrillard has defined identity as a 'brand of existence without quality' (Baudrillard 1999: 58). Not by chance, adolescents, as remarked by Lobet-Maris (2003), tend to base themselves on mobile brand names to belong to specific tribes, according to the well-known definition of Maffesoli (1988). The choice of a brand name as a way of defining mobile identities is also confirmed by other research (Danelon 2005).

Convergence

The last reason why the mobile phone is worthy of being described as the fourth communicative and cultural revolution is that it is the device in which the process of convergence is at this moment most significant. The significant and spontaneous convergence between orality and writing has matched voice and text, and now also image (Glotz *et al.* 2005). But it has also made possible the first forms of pre-interactivity with means of mass communication, which are heading towards the amplification – even if still only at the very beginnings – of the spaces of public debate. If the computer is a crossover between a typewriter, the postal system and an archive/library, the mobile phone is a crossover between the telephone, mobility and the intimate circle of relationships (Ito *et al.* 2005). But what is more important, the mobile has the premises to be the place where the convergence with the Internet and with television and radio take place. So, it is becoming both more multimedia and at the same time more humanized (Contarello *et al.* 2008). Of course new services and languages are needed to carry out this new stage, but already the mobile phone represents very well the idea of fusion which is the key principle of globalization (Goggin and Hjorth 2008; Hartmann *et al.* 2008; Nyiri 2008).

Let us focus on camera phones as the first example of mobile phone convergence which has introduced images (Hjorth 2007; Ito and Okabe 2006; Koskinen 2007). Of course now mobiles can produce, receive and send videos and receive television programmes. But the example of camera phones and simple pictures might be useful to outline a process of mobile communication which started with them. The hybridization of the human body by means of the mobile phone has led, together with the shift to camera phones and universal mobile telecommunications systems (UMTS), to a great development in virtualization, or rather, if I may borrow an expression from Baudrillard (1999), the construction of the double of the world. Reading the new technical developments of the mobile phone in this perspective may allow us to approach this new stage of the mobile phone with more suitable theoretical instruments. Generally the construction of a double of the world, as Baudrillard calls the virtual world, is attributed to the role played in the last ten years by the Internet. The significance of the mobile's contribution to the development of virtual reality is underestimated and misunderstood. In fact, the hybridization between the mobile phone and the human body has made virtualization of the real become not a generic doubling of the world, but the doubling of our personal shadow. Fortunati, in a study on mobile use by a group of adolescents, observed that they used the mobile to re-create a sort of virtual brotherhood and sisterhood (Fortunati 2001). In fact the use of the mobile by young people who are mainly single children and hence live on the family level in a world of practically only adults has the function of being able to build such close telephonic relations, out of reach of parents. These virtual relations act as calques of real relations between brothers and sisters. Wherever they are and whatever they do, even in the company of their parents, the adolescents are constantly in contact with their virtual brothers and sisters, and often start and end their day with a message or call. As we recalled earlier, Gergen, stressing the role of the mobile in constructing a special hybridized dimension made up of presence/absence, showed the intersection that is created with the use of the mobile in the public space, between the real world and its double (Gergen 2002). He showed the systematic upstaging of the real world in respect to the virtual one, the fact that the virtual one overshadows reality. But here it is worth asking ourselves why societies at a certain point put so much of their energies into the virtual world, also by means of the mobile, and have developed the technologies to make this a mass reality.

Through this double of the world, the political/economic system has been able to extend its dominion over another dimension of reality, recovering further spaces of action and volumes of exchange and communication as well as a new mode of command and control over labour, and being able to export many social tensions and conflicts. What is there, for instance, more controllable socially than a youth intent on playing a videogame or checking his/her e-mail or chatting? The fact is that on a general level, instead of acting politically to change the world, youth have preferred to abandon the real and to shelter in this pseudo-real world. Globalization has attempted to use the virtual world to its advantage and tried to beat other philosophies and more spontaneous, autonomous and libertarian approaches. Certainly the result of this conflict is still open. It can be seen that civil society has often been illumined by spontaneous movements that have organized forms of collective solidarity with the mobile (for example, on the occasion of the terrorist attack on the Russian school in Beslan, or of the 2004 Asian tsunami), but also by social movements that, as we recalled before, have organized forms of protest, public mobilization, the overthrow of political institutions (Nyiri 2003, 2005; Rheingold 2002), counter-information (Mblogs, for example), the possibility of controlling possible controllers and of photographing and recording their conduct.

Today, the mobile's construction of a double of the world has been endowed with technological convergence as it has acquired the image (camera phones), on-line chats and multimediality (Jenkins 2006). But what social significance does this desire and this practice of collecting images of the world assume? Is it perhaps another way to connect it to the virtual world, and to make reality disappear? Baudrillard suggests that photography is the kind of technique which, far from collecting the objectivity of the world, leads us into the heart of the false appearance of reality (Baudrillard 1999: 135). Taking a photograph means resisting the flux of everyday existence in the 360-degree space that surrounds us in order to mark a point and an instant, without however forgetting that the photographic image is a fiction, not a representation. But photographing an object also means registering the absence and concealment of the object from the gaze of the subject: in fact what remains is never the 'Other', but what remains of the 'Other' when the 'Other' is not there (ibid.: 138).

What are camera phones able to produce through this kind of symbolic murder that takes place in the act of photography and in which both the object and the subject disappear? The research conducted so far on multimedia messaging services speaks of a 'banalization' of the everyday (Koskinen 2007). It seems to be iconic chatter which, in the words of Bachelard, acts as eternalization of 'unconsciousness archetypes' and which, according to Maldonado (2005), occurs as part of a reduced repertory of pseudo-significant nuclei that are constantly being repeated and crystallized in the act. But, for Baudrillard, there is more to it than that: 'We complain that the real has disappeared with the pretext that everything passes through the image. But we forget that the image itself disappears under the blows of reality' (Baudrillard 1999: 139). And that it is instead necessary to free the image from the real, thus restoring the specificity and power to the image and freeing the real from its reality principle (ibid.: 140). The photographic image shows itself then as revealing what is undecipherable and incommunicable in each of us. The photograph of the everyday that the camera phone takes is the ideal instrument to reaffirm the counterfeiting of the real that continues to take place under globalization – the concealment of reproduction as the sphere of value production. As a photograph does not analyse reality, but rather places itself literally on the surface of things and illustrates their apparition in the form of fragments, and for a very short length of time (ibid.: 136–37), it seems to be the ideal technique for recoding the domestic sphere and the everyday, thus reaffirming its 'disvalue'.

This urge to observe, produce and preserve the images of the world through the mobile phone is clearly related to the general shift that has taken place from the material dimension to the immaterial. The immateriality of the image in fact does not involve only its double life, but also a different intensity of the material (much less pronounced), which therefore leads us to a world of insufficiency of the existent. At the same time, if observed from another point of view, camera phones offer a new possibility for building a social memory of everyday life (Green 2009) and becoming a powerful new research tool. Under globalization, it is for the precise reason that the mobile has been centred from the start on the individual that it has been able to keep up the process of fusion in all its manifestations, appearing as the symbol of post-modernity. I am thinking of the boundaries between the public and private space that have been merged and redrawn many times, together with the rules of good manners; or the dimension of intimacy which has hybridized with that of extraneousness; or again the dimension of mobility that has merged with that of sedentariness – the mobile phone used very much at home, at work, etc.; or the hybridization of the mobile with fashion; or the capacity of the mobile to interpret the needs of flexibility, individuality and mobility in post-Fordist society. And the list could go on. The mobile phone has acted as a catalyst for many aspects of the fusion process and has represented them socially and made them evident. Apart from photography, the mobile has also shown a great capacity for structural convergence with other technologies and a great capacity to absorb trans-medial content such as information, music and publicity.

The mobile has extended the interactive qualities of the telephone in respect to the also-powerful electronic mass media – radio and TV. More specifically, it has not only made contact possible with other means of communication in real time, even when on the move, via direct telephone calls, but it has also offered the possibility of simple unidirectional interaction through texting, which originated the practice of voting on radio and TV programmes and the various TV and radio transmission sites on the web. The ever-more common practice of the electronic vote has in its turn developed through the implementation of special software for the rapid elaboration of the voting data.

Writing, which had been completely expelled from the radio and almost completely from TV (I say 'almost' because actually the TV screen is becoming more and more written, under the influence of the computer and above all the Internet), has been snatched back by mobile users within a dimension of minimalism. The higher cost of phone calls in comparison to SMS has forced users to reduce communication time, in this way leading to a prevalence of rapid conversation or more often brief messages. Tied to the organization of daily life, they have, however, had a great phatic function as well as an anti-anxiety function (which in its worst forms can produce a pathological dependence on the mobile phone at the communicative level – see Vincent 2003). Mobile writing, which is quite hybridized with orality, has aimed here at celebrating the needs of the organization of daily life, the aphoristic message, the quick agreement of appointments, and the like. It is a writing that, as I have said before, is casual, and has the same antiformal spirit as a T-shirt or a pair of jeans. We may say that a text message can be considered the emblem of the powerful processes of simplification that have overtaken writing (Baron 2008), which was humanity's first important communicative technology, and is probably one of the channels in which the dissolution of meaning, which is attacking the society of knowledge, is most manifest.

Conclusion

In this chapter I hope to have managed to reconstruct the main social phenomena and processes illustrating how the mobile phone has played a specific role in the shift to the fourth communicative and cultural revolution, and which coincides with the shift to post-Fordism and post-modernity. I have shown the ways in which this small instrument has accompanied and enhanced such radical changes. It is the technology which has had the fastest and widest global diffusion. This means that this device is considered useful and convenient by users and potential users in societies at different stages of industrialization. I have shown that the mobile phone is at the core of globalization, for its usability by both the literate and illiterate, for its low costs (at least regarding SMS), and for its great adaptability, which allow it to be reinvented by different users and used in different ways and for various purposes. After all, the speed of the diffusion of the mobile phone reflects its importance to users, probably because of the margin of freedom that it offers and also because of the new level of democratization that results from this device. The use of the mobile phone, being personal, strengthens the power and the control of the individual on the machine and on the communication mediated by it, reinforcing the most powerful social process in capitalist societies: individualization. Furthermore, by reducing the distance from the human body, the mobile has begun the process of hybridization between ICTs and the human body, opening a new ontology of the human being. Finally, the mobile phone is the technology most probably destined to include a plurality of tasks. It is not by chance that convergence is waging its greatest battle on the mobile phone.

It is hoped that in the future it will be possible to produce further empirical studies and theoretical reflections to defend this thesis. The direction in which we should multiply our efforts is that of understanding better not only the social role of the mobile phone but also the general context (post-modernity, globalization) in which this device operates.

Note

1 The first version of this chapter was presented at the international workshop 'Technology, Time and Everyday Life' organized by the Oxford Internet Institute, Friday 25 November 2005.

References

Appadurai, A. (1990) 'Disjuncture and difference in the global cultural economy', in M. Featherstone (ed.) *Global Culture: nationalism, globalization and modernity*, London: Sage.

Aronson, S. (1977) 'Bell's electrical toy: what's the use? The sociology of early telephone usage', in I. De Sola Pool (ed.) *The Social Impact of the Telephone*, Cambridge, MA: MIT Press.

Baron, N. (2008) 'Adjusting the volume: technology and multitasking in discourse control', in J. Katz (ed.) *Handbook of Mobile Communication Studies*, Cambridge, MA: MIT Press.

Baudrillard, J. (1999) *L'échange impossible*, Paris: Éd. Galilée.

Beniger, J. (1986) *The Control Revolution: technological and economic origins of the information society*, Cambridge, MA: Harvester.

Castells, M., Fernandez-Ardevol, M., Qiu, J.L. and Sey, A. (2004) 'The mobile communication society: a cross-cultural analysis of available evidence on the social uses of wireless communication technology', paper presented at the International Workshop on Wireless Communication Policies and Prospects: A Global Perspective, Annenberg School for Communication, University of Southern California, Los Angeles, October.

—— (2007) *Mobile Communication and Society: a global perspective*, Cambridge, MA: MIT Press.

Contarello, A., Fortunati, L., Gomez Fernandez, P., Mante-Meijer, E., Vershinskaya, O. and Volovici, D. (2008) 'ICTs and the human body: an empirical study in five countries', in E. Loos, L. Haddon and E. Mante-Meijer (eds) *The Social Dynamics of Information and Communication Technology*, Farnham: Ashgate.

Danelon, E. (2005) *Il cellulare: solo un mezzo di comunicazione?* Tesi di laurea: relatore, prof. Leopoldina Fortunati.

Donner, J. (2008) 'Research approaches to mobile use in the developing world: a review of the literature', *The Information Society*, 24(3): 140–59.

Fortunati, L. (2001) 'The mobile phone between orality and writing', in ICUST, *E-usages*, Paris: France Telecom.

—— (2007) 'Immaterial labor and its machinization', *Ephemera: Theory & Politics in Organization*, 7(1): 139–57.

Fortunati, L., Katz, J. and Riccini, R. (eds) (2003) *Mediating the Human Body: technology, communication and fashion*, Mahwah, NJ: Erlbaum.

Gebhardt, J. (2010) 'Alfred Schütz and the media: the inter-subjective constitution of mediated interpersonal communication in everyday life', in L. Fortunati *et al.* (eds) *Interacting with Broadband Society*, Berlin: Peter Lang.

Gergen, K. (2002) 'The challenge of absent presence', in J. Katz and M. Aakhus (eds) *Perpetual Contact: mobile communication, private talk, public performance*, Cambridge: Cambridge University Press.

Gershenfeld, N. (1999) *When Things Start to Think*, New York: Henry Holt and Company.

Glotz, P., Bertschi, S. and Locke, C. (eds) (2005) *Thumb Culture: the meaning of mobile phones for society*, Bielefeld: Transcript Verlag.

Goggin, G. and Hjorth, L. (eds) (2008) *Mobile Technologies: from telecommunications to media*, London: Routledge.

Green, N. (2009) 'Mobility, memory and identity', in G. Goggin and L. Hjorth (eds) *Mobile Technologies: from telecommunications to media*, New York: Routledge.

Griswold, W. (1994) *Cultures and Societies in a Changing World*, Thousand Oaks, CA: Pine Forge Press.

Habermas, J. (1989) *The Structural Transformation of the Public Sphere*, Cambridge: Polity.

Haddon, L. (2004) *Information and Communication Technologies in Everyday Life: a concise introduction and research guide*, Oxford: Berg.

Hartmann, M., Rossler, P. and Höflich, J. (eds) (2008) *After the Mobile Phone: social changes and the development of mobile communication*, Berlin: Frank & Timme.
Harvey, D. (1989) *The Condition of Postmodernity*, Oxford: Blackwell.
Hjorth, L. (2007) 'Engagement rings: a cross-cultural analysis of camera phone genres, modes of sharing and digital storytelling', paper presented at The Future of Digital Media Culture: 7th International Digital Arts and Culture (DAC) Conference, Perth, Australia, September.
Ito, M. and Okabe, D. (2006) 'Everyday contexts of camera phone use: steps towards technosocial ethnographic frameworks', in J. Höflich and M. Hartmann (eds) *Mobile Communication in Everyday Life: an ethnographic view*, Berlin: Frank & Timme.
Ito, M., Okabe, D. and Matsuda, M. (eds) (2005) *Personal, Portable, Pedestrian: mobile phones in Japanese life*, Cambridge, MA: MIT Press.
Jameson, F. (1991) *Postmodernism, or the Cultural Logic of Late Capitalism*, Durham, NC: Duke University Press.
Jenkins, H. (2006) *Convergence Culture: where old and new media collide*, New York: New York University Press.
Katz, J. (ed.) (2003) *Machines that Become Us*, New Brunswick, NJ: Transaction.
Katz, J. and Aakhus, M. (2002) *Perpetual Contact: mobile communication, private talk, public performance*, Cambridge: Cambridge University Press.
Koskinen, I. (2007) 'Managing banality in mobile multimedia', in R. Pertierra (ed.) *The Social Construction and Usage of Communication Technologies: European and Asian experiences*, Singapore: Singapore University Press.
Licoppe, C. and Smoreda, Z. (2005) 'Rhythms and ties: towards a pragmatics of technologically-mediated sociability', in R. Kraut, M. Brynin and S. Kiesler (eds) *Domesticating Information Technologies*, Oxford: Oxford University Press.
Ling, R. (2004) *The Mobile Connection: the cell phone's impact on society*, San Francisco, CA: Morgan Kaufmann.
Ling, R. and Pedersen, E. (eds) (2005) *Mobile Communication and the Re-negotiation of the Social Sphere*, London: Springer.
Liu, G. (2005) 'The use of ICTs in disseminating sex service information: impacts upon one-woman brothel workers in Hong Kong and some of their responses', paper presented at Mobile Communication and Asian Modernities II: Information, Communications Tools & Social Changes in Asia, Beijing, China, October.
Lobet-Maris, C. (2003) 'Mobile phone tribes: youth and social identity', in L. Fortunati, J. Katz and R. Riccini (eds) *Mediating the Human Body: technology, communication and fashion*, Mahwah, NJ: Erlbaum.
Lyotard, J-F. (1979) *La condition postmoderne: rapport sur le savoir*, Paris: Minuit.
Maffesoli, M. (1988) *Le Temps des tribus*, Paris: Le Livre de Poche.
Maldonado, T. (1997) *Critica della Ragione Informatica*, Milan: Feltrinelli.
—— (2005) *Memoria e Conoscenza: sulle sorti del sapere nella prospettiva digitale*, Milan: Feltrinelli.
Marx, K. (1857–58) *Grundrisse der Kritik der politischen Ökonomie* (Robentwurf); trans. It. (1968) *Lineamenti fondamentali della critica dell'economia politica*, 2 vols, Florence: La Nuova Italia.
Myerson, G. (2001) *Heidegger, Habermas and the Mobile Phone*, Cambridge: Icon Books.
Nyiri, K. (ed.) (2003) *Mobile Democracy: essays on society, self and politics*, Vienna: Passagen Verlag.
—— (ed.) (2005) *A Sense of Place: the global and the local in mobile communication*, Vienna: Passagen Verlag.

—— (ed.) (2008) *Integration and Ubiquity: towards a philosophy of telecommunications convergence*, Vienna: Passagen Verlag.

Oksman, V. and Rautiainen, P. (2003) 'Extension of the hand: children's and teenagers' relationship with the mobile phone in Finland', in L. Fortunati, J. Katz and R. Riccini (eds) *Mediating the Human Body: technology, communication and fashion*, Mahwah, NJ: Erlbaum.

Rheingold, H. (2002) *Smart Mobs: the next social revolution*, New York: Perseus.

Schütz, A. (1967) *The Phenomenology of the Social World*, Evanston, IL: Northwestern University Press.

Sennett, R. (1977) *The Fall of Public Man – on the social psychology of capitalism*, New York: Alfred A. Knopf.

Sørensen, C. and Pica, D. (2005) 'Tales from the police: rhythms of interaction with mobile technologies', *Information and Organization*, 15: 125–49.

Spivak, G. (1999) *A Critique of Post-Colonial Reason: toward a history of the vanishing present*, Cambridge, MA: Harvard University Press.

Tomlinson, J. (1999) *Globalization and Culture*, Cambridge: Polity Press.

Van Dijk, J. (1999) *The Networked Society: an introduction to the social aspects of new media*, London: Sage.

Vincent, J. (2003) 'Emotions and mobile phones', in K. Nyiri (ed.) *Mobile Democracy: essays on society, self and politics*, Vienna: Passagen Verlag.

Wajcman, J. (2004) *TechnoFeminism*, Cambridge: Polity Press.

Wearing, S. (2002) *Mobiles a Confidence Booster for Young Women*. Online. Available: www.uts.edu.au/new/releases/2002/July/24.html (accessed 8 January 2010).

Werner, C. and Haggard, L. (1985) 'Temporal qualities of interpersonal relationships', in M. Knapp and J. Daly (eds) *Handbook of Interpersonal Communication*, London: Sage.

Williams, F. (1982) *The Communications Revolution*, Beverly Hills, CA: Sage.

5 The history of the mobile phone in China

Jing Wang and Chung-tai Cheng

Introduction

China has experienced rapid development in the field of mobile phones. This chapter traces the progress of this development, which is divided into four stages:

- 1987–1993 – mobile phones as a luxury item;
- 1994–1998 – consumption of mobile phones amongst high-income earners;
- 1999–2004 – the spread of mobile phone consumption to the populace at large;
- 2005–2010 – the high prevalence of mobile phone usage and the diversification of the mobile phone market.

The factors that account for the development of mobile phone use include the improved living conditions of Chinese people, advances in the Chinese telecommunication industry, the effects of government policies and the flourishing of domestic mobile innovation, and changes in the mobile communication infrastructure which support a growing number of mobile phone users.

Four stages of development of mobile phone use in China

1987–1993 – mobile phones as a luxury item

China began its era of mobile communication in the late 1980s. In November 1987, cellular phones on the analogue standard of 900 MHz TACS (Total Access Communication Standard) were first introduced into Guangzhou, the capital city of Guangdong province (Wang 1994; Yao 1999). However, because of technical limitations with the mobile communications network and the immaturity of the technology associated with the production of the mobile phone, mobile users had to spend nearly 30,000 RMB in order to buy a mobile phone and pay an additional network access fee of over 10,000 RMB, not including expensive service charges. At that time, the per capita GDP in China was only 1,140 RMB: the average money wage was 1,459 RMB per year (State Statistical Bureau 1988). It meant that, if one wanted to buy a mobile phone, one would have to save for at least 30 years. A mobile phone even cost the same as a Toyota agricultural vehicle in comparison with other consumer goods (Wu 2005).

As one of the first mobile phone consumers, Xu Feng, who is now the president of Guangdong Zhonghai Corporation, spent 20,000 RMB to buy a mobile phone with the help of his relatives in 1987. At that time, there was no established price when the first sets of mobile phones were introduced in China. Mr Xu recalled:

> People who were in charge of these phones in the bureau of post and telecommunications didn't know how much to sell them for. They just asked me to leave a check for 20,000 RMB, in order to take away the first phone.
>
> (Wu 2005)

The Sixth National Games, held in Guangzhou in 1987, provided an opportunity to promote the introduction of the first 100 sets of analogue cellular mobile phones in China. However, only one in three urban dwellers owned a colour TV set, only one in five owned a refrigerator and as few as one in seven owned a camera during that period (State Statistical Bureau 1988). A mobile phone represented an unobtainable luxury for the bulk of the population at the low level of household spending. Even if one had enough money to buy one, the process of buying and repairing it was quite complicated. Mr Xu recalled that prospective users had to invite mobile phone dealers to dinner to establish *guanxi* (interpersonal relationships and connections) so as to make it easier to buy a mobile phone and repair it when needed. Moreover, buyers and those who repaired the phones had to deal with long queues, impatient and unpleasant sales personnel and the complex procedure needed to apply for the phone, followed by long periods of waiting for approval. At the same time, the illegal mobile phone market became so rampant in Guangzhou that a mobile phone with a number could be sold at 60,000 RMB on Lingyuan West Road, an area with a concentration of telecommunication products in Guangzhou (Wu 2005). Nevertheless, since Guangzhou was the only place where mobile phones could be used because of the limited network, the cost of maintaining the phone could be over 10,000 RMB a month.

In those early years, mobile phone users in China could be roughly divided into three categories: successful businessmen like Xu Feng, high-ranking government officials and those in the special forces such as the army. Their phones were analogue with phone numbers that comprised six digits from 901001 to 901099. For example, Xu Feng's number was 901088; it was an NEC product with an antenna which had to be extracted in order to make a call (Wu 2005). Although there were just 300 mobile phone users all over China in 1987, the number of users rose to 3,100 in 1988, increased to 18,000 in 1990 (Yao 1999) and rose substantially thereafter (Table 5.1). By 2001, the mobile phone consumption in China was approaching 150 million (MIIT 2001). There was no communication product which could compare with the rate at which mobile phone use swept across China.

Table 5.1 Number of mobile phone users (1987–2001)

Year	*Number of mobile phone users*
1987	300
1988	3,100
1990	18,000
1992	176,900
1994	1,567,800
1997	13,232,900
2001	145,222,000

Sources: State Statistical Bureau, *China Statistical Yearbook*, various years, China Statistics Press; Ministry of Information Industry Technology, The People's Republic of China, various years.

In 1992, although the number of users was confirmed at 176,900 (State Statistical Bureau 1995), the price of a mobile phone was still too high for most consumers. The purchase of a phone with a number cost 15,000 RMB, in addition to a network access fee of 6,000 RMB, an instalment fee of 460 RMB, and 150 RMB in monthly rental. These costs did not include the cost of conversing at 0.5 RMB per minute (Yi 2005). Although per capita GDP in China had doubled from what it had been in 1987 to 2,287 RMB in 1992 per head (Table 5.2) and consumption of household goods had increased whereby nearly 75 per cent (74.87 in every 100) of urban residents owned a colour TV, one in two urban people owned a refrigerator, and one in four owned a camera (State Statistical Bureau 1993), owning a mobile phone was still beyond the reach of the majority of people, who would have had to save for ten years to be able to afford one.

1994–1998 – consumption of mobile phones amongst high-income earners

By 1994, the penetration of the mobile phone had broken through the 1 million mark, reaching 1.568 million. From this time onwards, the number of users almost doubled every year, and this level of growth at over 80 per cent per annum did not slow down until 2001 (State Statistical Bureau 2002). There are two main reasons for this escalating rate of growth: one is the constant increase in household income, which stimulates the quantity of demand of mobile phones; another one is the end of the monopoly in the supply market of mobile phones, which contributes to serious competition and technical improvements in quality.

Before 1994, postal and telecommunication services were under the administration of a single department, the Ministry of Post and Telecommunications (MPT), which was established by the State Council in 1979. Local post and telecom enterprises were under the leadership of both the MPT and provincial governments, which were directly controlled by the MPT. In this circumstance, mobile communications were monopolized by the MPT and as a result the installation of a fixed-line phone cost as much as 2,000 RMB. It has since fallen to its

Table 5.2 Consumption of household goods in 1987 and 1992

Year	*Per capita GDP (yuan)*	*Average money wage (yuan)*	*Spread rate of colour TVs in urban areas*	*Spread rate of refrigerators in urban areas*	*Spread rate of cameras in urban areas*
1987	1,104	1,459	34.63	19.91	14.34
1992	2,287	2,711	74.87	52.60	24.32

Source: State Statistical Bureau, *China Statistical Yearbook* (1988, 1993), China Statistics Press.

current level of 200 RMB. The waiting time for installation of a fixed-line phone was nearly a month, and the service was poor. Procuring a mobile was even more difficult, for only in certain appointed places could one buy a mobile phone with network access. In addition, surcharges, which included the network access fee, were often two or three times more than the price of a mobile (Dong and Song 2000). Therefore, mobile phone producers had to pay much attention to relationships with the officials of the MPT and with the provincial government because they were required to obtain permission for network access from the MPT if they wanted to conduct any mobile phone business.

In order to improve the market competitiveness of the telecommunication business and enter the WTO, the Ministry of Electronics Industry (MEI) joined with the Ministry of Energy, the Ministry of Communication, the Ministry of Water Resources, the Ministry of Electric Power and the Ministry of Petroleum, to form the 'China United Communications Company' (China Unicom) on 19 July 1994. Its major business was, and still is, mobile communications. This merger ended the MPT's monopoly. In 1998, another important development took place: the Ministry of Information Industry (MII) was formed on the basis of the merger of the MPT and MEI. After that the postal and telecommunication services were separated and restructured, which led to the establishment of two independently operated departments, China Post and China Telecommunications, which had previously been a single entity. A year later, in 1999, the State Council considered a plan to divide China Telecommunications into four separate commercial entities – fixed-line communications, mobile communications, paging, and satellite communications, which gave rise to China Telecommunications Group Company (China Telecom), China Mobile Communications Group Company (China Mobile), China Paging Communications Group Company (which was then taken over by China Unicom) and China Satellite Communications Group Company (which was launched in 2001). China Telecom took over fixed-line services, while China Mobile handled mobile phone business. In May 2000, China Mobile acquired China Telecom Group's market share of China Telecom (HK) Company Ltd at no cost, and eventually became the new owner of China Telecom (HK), which then changed its name to China Mobile (HK). The two companies which provided mobile communication services – China Unicom and China Mobile – started to operate at full capacity. During this process, China Netcom and China Railcom were established by the MII – the former was established in 1999, the latter in 2001. They both deal in part with fixed-line business and in part with Internet access. Then, in November 2001, China Telecom, which monopolized the nation's fixed-line business, was divided into two parts along regional lines – the new China Telecom Corporation and the new China Netcom Communications Group Corporation. The newly merged China Netcom took over the fixed-line business in ten provinces in north China, while the reborn China Telecom inherited the rest in the south. After the appearance of the two giant carriers of fixed lines, the market for telecommunications was formed. The six players in the market were China Unicom, China Mobile, China Telecom, China Netcom, China Railcom and China Satcom (Lu and Wong 2003). Since then, for mobile network services, China Unicom and China Mobile have competed with each other incessantly. China Mobile took up the GSM (Global System of Mobile Communications) before the turn of the century, a time when mobile phones could almost be used all over China. China Mobile has provided a number of popular services such as '*quanqiutong*' (GoTong), '*ShenzhouXing*' and '*donggandidai*' (M-zone). All users whose mobile phone numbers begin with 134 to 139, 150, 151, 158 and 159 belong to China Mobile, some of which are approved for third generation (3G) mobile phones. As for China Unicom, it is the sole operator of CDMA (Code Division Multiple Access) authorized by the State Council. Because CDMA, a sophisticated technique developed in the United States, has features including a high-quality signal, rapid connecting rates, good voice quality, effective security and compatibility with the analogue network, CDMA has become one of China Unicom's main means of competing against China Mobile's GSM network.

Domestic reform was accomplished by the gradual entry of foreign mobile phone companies in China. As early as 1992, the Motorola Company established its first base in Tianjin – Motorola (China) Electronics Corporation Ltd. It invested US$90 million in setting up a second base in Suzhou in 1997. By 2000, Motorola investment in China had reached US$2 billion. Another example is Ericsson. In 1995, Ericsson established a joint venture with the Head Office of China Post and Telecommunications Industry Corporation and Hong Kong Wing Sun Enterprise. The joint company was called Beijing Ericsson Mobile Communications Corporation. Thereafter it set up another joint venture with Panda Electronics Group – Nanjing Ericsson Mobile Communications Corporation. In April 1995, Nokia also set up a joint venture with the Head Office of China Post and Telecommunications Industry Corporation and Beijing Telecommunications Equipment Plant of the MPT to set up Beijing Nokia Mobile Communications Corporation Ltd. In May 1995, it established the Dongguan Nokia Mobile Phones Corporation Ltd in conjunction with the Communications Development Corporation of Dongguan. Subsequently, Siemens, Philips and others entered the Chinese market one after another to start joint ventures (Yao 1999; Zhu and Chen 1999).

According to a survey conducted amongst urban residents in China's main cities in 1998 (Zhu and Chen 1999), the mobile phone market was dominated by foreign brands, with Motorola claiming 37.3 per cent, Ericsson 28.6 per cent and Nokia 15.6 per cent, with the remaining 20 per cent shared by Philips, Siemens, Alcatel, Sony and others. Competition amongst well-known international mobile

Table 5.3 Consumption of household goods (1994–1998)

Year	*Per capita GDP (yuan)*	*Average money wage (yuan)*	*Spread rate of colour TVs in urban areas*	*Spread rate of refrigerators in urban areas*	*Spread rate of cameras in urban areas*
1994	3,923	4,538	86.21	62.10	29.83
1995	4,854	5,500	89.79	66.22	30.56
1996	5,576	6,210	93.50	69.67	32.13
1997	6,053	6,470	100.48	72.98	33.64
1998	6,307	7,479	105.43	76.08	36.26

Source: State Statistical Bureau, *China Statistical Yearbook*, various years, China Statistics Press.

phone corporations not only contributed to improvements in mobile technologies but also a considerable drop in the purchase cost of the phones. Therefore, the number of mobile phone users increased substantially, that is, in 1997 the number of users reached 13.233 million and in 1999 this jumped to 40 million. China, by then, had become the world's second-largest mobile phone market, second only to the United States (Dong and Song 2000; MIIT 1997, 1999).

The burgeoning demand for mobile phones was abetted by China's steadily increasing per capita GDP as well as increasing average personal wages and per capita annual income (see Table 5.3). By 1998, these three figures had increased fourfold since the inception of the mobile phone 11 years earlier. As living standards rose, the number of mobile phone owners increased. Furthermore, consumption of colour TVs rose to more than one per urban household and urban dwellers began to purchase video cameras and personal computers, and developed an interest in surfing the Internet. As living standards rose, the number of mobile phone owners increased. By 1997, the consumption of mobile phones in urban areas had reached 1.7 sets per 100 people (State Statistical Bureau 1988).

As mentioned above, improvements to the telecommunications infrastructure and the fierce international competition in the Chinese mobile communication market brought down the price of mobile phones while related fees also decreased substantially. Between 1994 and 1999 the price of a mobile phone dropped by some two-thirds from over 10,000 RMB to 3,000–4,000 RMB; and the network access fee, as well as some service fees, also fell by 70 per cent between those years (Dong and Song 2000).

1999–2004 – the spread of mobile phone consumption to the populace at large

During the period between the late 1980s and early 1990s, only the wealthy or senior officials had the means and affordability to purchase a mobile phone. Since 1994, the middle class joined this group of consumers, and from 1999 onwards, mobiles became popular amongst the populace as a whole as the price of mobiles began to fall at the end of 1998. Because of severe market competition, the price of popular mobile phone brands decreased significantly. For instance, the price of Motorola, Ericsson and Nokia products fell by more than 50 per cent around late 1998 to early 1999. For example, the Motorola GC87C, which cost over 5,000 RMB when it first appeared on the market in 1997, was sold at a price of 800 RMB in 1999. The Motorola CD928+ dropped by 1,000 RMB in less than two months, since its first appearance in late 1998. The price of the Nokia 8810, which had been fixed at 16,000 RMB when it came onto the market, dropped by 50 per cent to 8,000 RMB over a period of six months. In 1999 prices continued to fall: the price of a Nokia 8810 was cut by another 50 per cent and it sold for 4,000 RMB in June 1999. The Nokia 5110, which was aimed at the lower end of the market, also fell in price from 3,000 RMB to 1,500 RMB from 1998 to 1999 (*Jisuanji Shijie* [Computer World] 2000). Notably, these dramatic price decreases could be attributed to technological advances that led to lower costs and higher profits, which in turn made the price of the phones more accessible to a growing number of consumers and thus the market share widened. As a result, more and more companies joined the line of mobile phone producers.

Some domestic enterprises and scientific research institutions began to innovate and develop domestic mobile phones and networks, with the support of the government's policies and the opportunity of China's WTO entry. When China signed up to the World Trade Organization (WTO) in November 1999, it committed itself to allowing up to 49 per cent foreign ownership and control in basic telecom service ventures and 50 per cent in value-added services after its accession to the WTO according to an agreed timetable. As a result competition would be fiercer and strategies would be required to protect the national industry. In December 2001, China officially became a member of the WTO (Lu and Wong 2003).

On the other hand, Datang, Huawei, Zhongxing and others succeeded one after the other in obtaining network access licences for telecommunications equipment production from the MII after 1999 (Dong and Song 2000). For example, 'Kejian', which had connections with the Chinese Academy of Sciences, engaged in the production and consumption of mobile phones. TCL, Haier, Konka (*Kangjia*) and Lenovo (*Lianxiang*), which traditionally dealt in electrical equipment, also entered the mobile phone industry (Yi 2005). In May 1999, Guangzhou South High-tech Corporation Ltd was established on the basis of the seventh research institute of the MII, which dealt, in particular, with the research, development, production and consumption of domestic mobiles.

Although there were many debates about whether or not China should develop mobile phones domestically, the domestic brands flourished and therefore the government created new systems to regulate and protect mobile phone productions in China. On 31 December 1998, the MII issued the regulation 'Arrangements for the Approval of Network Access of Telecommunications Equipment' to be effective from 1 January 1999. Under the approved regulation, all telecom equipment to be used in public or private telecom networks in China was compelled to operate under a Network Access Licence (NAL) and a Network Access Identifier (NAI) which were to be obtained from the MII (Lu and Wong 2003). The aim of this regulation was to restrict the entry of foreign brands onto the Chinese market in order to protect the national telecommunication industry and maintain a certain degree of market control. Afterwards, the same 'licence' regulation system (*pai-zhao zhidu*) was extended to the field of mobile phone production when domestic mobile phones became the focus of investment (Xu 2005). In 2006, the MII had issued 49 of these 'licences', including 30 for GSM and 19 for CDMA. There were 14 companies which had obtained both GSM and CDMA licences. Among the companies which had GSM licences, there were 13 joint ventures and 17 domestic companies; and among the 19 CDMA licences, only one did not belong to a domestic company, having been obtained by Motorola (Peng 2005).

In 2000, the government invested 1.4 billion RMB from network access fees to support the development of domestic mobile phones in 2000. For five successive years, the government decided to draw 5 per cent from installation fees obtained from fixed lines, as a research fee for the innovation and production of domestic mobile phones (Bi 2000). After that, domestic mobile phones started to capture the consumer market gradually, especially at the lower end. In the same year, domestic mobile phones held 8 per cent of the market, increasing to 15 per cent in 2001, nearly 30 per cent in 2002, and 60 per cent in 2003. In comparison, the foreign mobile phone alliance, composed of Motorola, Nokia, Siemens and Ericsson, held just 24 per cent of the market, having declined from 84 per cent in 1999 (Qian and Yang 2004). Moreover, as one of the most popular domestic brands in China, Bird is known as the 'fighter among mobile phones' because of its advanced technicality. Connections with a Bird mobile phone are exceptionally good because its partner is Sagem of France, famous for its air force communications. This domestic brand overtook Motorola as the biggest player in the domestic market for mobile phones in 2003, with a 15.01 per cent share. Meanwhile, aiming at focusing on the research and development of technology, another local brand, TCL, invested an immense amount of money in recruiting research talent domestically and overseas and established a TCL research institute for mobile communications which employed hundreds of PhD and Master's Degree graduates. In addition it established research centres and it co-operates with research institutes in America, Japan, France, Korea and elsewhere. TCL created mobile phones that were especially suited to the small thin hands of Chinese people. Through establishing a distinctive local brand culture which comprised jewelled decorations which symbolized luck, honour and friendship, TCL overtook Nokia as the third biggest player (Hu 2003; MIIT 2003).

On 1 June 2001, all 17 domestic mobile phone companies set up the '*guochan shouji qiye shounao luntan*' ['The forum of the heads of domestic mobile phone enterprises']. Its purpose was to combine the strengths of these enterprises to improve their domestic competitiveness, and to promote the use of domestic mobile phones and the development of the production of domestic mobile phones in order to boost mobile communications in China (Zhang 2001). In spite of these efforts, over 50 per cent of domestic mobile phones are still produced through putting together parts produced by other producers and labelling them with domestic brand labels. Up to

2001, domestic mobile phones which relied on their own technologies accounted for just 30 per cent of the domestic market, while the remaining 70 per cent of the domestic market was held by other technologies using other parts (Xiang 2001).

As the network access fee was cut and the price of mobile phones continuously decreased, the number of mobile phone users increased sharply. By the end of 1999, the number had reached 43,296,000, which made China the second-largest mobile phone market in the world (MIIT 2000). As early as July 1999, it was expected that existing mobile phones would not be sufficient to meet the demand from the increasing number of users. As a result, the number of digits needed for World Tone users rose from 10 to 11, with the users adding a '0' after the third digit (*Yidongju* [Mobile Bureau] 1998). On 1 July 2001, the MII and Ministry of Finance jointly abolished both the installation fee for fixed telephone lines and the network access fee for mobile phones (Yu *et al.* 2001). In August 2001, China overtook the United States as the world's largest mobile phone market when the number of users hit 120.6 million, compared with 120.1 million in the United States (People 2001).

From 1999 to 2002, the spread rate of mobile phones in China increased from 3.5 per cent to 16.1 per cent, and in urban areas the rate reached 62.89 per cent in 2002, compared with 7.14 per cent in 1999 (State Statistical Bureau 2004). In October 2003, the number of mobile phone users in China had begun to exceed the number of fixed-line users, with 256.94 million mobile users and 255.14 million fixed-line users. By the end of 2003, the number of mobile users was 268.69 million, 30 per cent more than it was in 2002, as a result of average growth of 52,240 users every month (MIIT 2003). At this time, one out of five people had a mobile phone in China, and in the urban areas 9 out of 10 people had a mobile phone (State Statistical Bureau 2004). By the time the per capita GDP had reached 9,101 RMB, every urban person owned 1.31 TV sets on average, and there was one computer for every 27.81 people (State Statistical Bureau 2004). At the same time, the number of people who surfed the Internet had increased to 79.5 million, that is, 130 times more than six years earlier and 9 times more than there had been in 1999 (Information Center of China Network 2003). (See Table 5.4.)

In the meantime, there was also a boom in the use of the short message service (SMS) from 2000. According to data provided by China Mobile, the number of SMS messages sent through China Mobile in 2000 was 1 billion and it reached 15.9 billion in 2001. The 90 billion messages sent in China in 2002, including 15 billion sent by users of China Unicom (Kuang 2003), represented a significant share of the world's total in 2002 of 360 billion messages. In other words, for every four messages sent throughout the world in 2002, one was sent by a person in China. In 2003, the number of messages sent in China hit 150 billion (Yin *et al.* 2004). According to a report quoted in *Beijing Daily* on 13 February 2003, the number of New Year greetings was 7 billion during the first few days of the Spring Festival, which means each user sending more than 30 messages during these days. Roughly speaking, one SMS message costs 0.1 RMB for non-package service plans, so that 150 billion messages yield 15 billion RMB income revenue to mobile service providers. For Sina (www.sina.com), Sohu (www.sohu.com) and NetEase (www.163.com), the three most popular websites in China, their SMSs

Table 5.4 Consumption of household goods (1999–2004)

Year	*Per capita GDP (yuan)*	*Average money wage (yuan)*	*Spread rate of video cameras in urban areas (per household)*	*Spread rate of computers in urban areas (per household)*	*Spread rate of mobile phones in urban areas (per household)*	*The number of Internet users (million)*
1999	6,534	8,346	1.06	5.91	7.14	8.9
2000	7,086	9,371	1.30	9.70	19.50	22.5
2001	7,651	10,870	1.60	13.30	34.00	33.7
2002	8,214	12,422	1.92	20.63	62.89	59.1
2003	9,101	14,040	2.45	27.81	90.07	79.5
2004	10,561	16,024	3.17	—	111.35	94.0

Sources: State Statistical Bureau, *China Statistical Yearbook*, various years, China Statistical Press; the China Internet Network Information Center, The People's Republic of China, various years.

generated more than 30 per cent of the total revenue in 2003 (Kuang 2003). The reason for such rapid development of the SMS in China can be traced back to 1999. At that time, Nokia, Ericsson, Siemens, NEC and other companies successively introduced mobile phones equipped to provide Chinese characters, that is, there are two input methods commonly used to insert Chinese characters: 'pinyin' (which is based on the pinyin method of romanization) and 'bihua' (which is based on the stroke orders of Chinese characters). Domestic companies, such as Kokan, Amoi and Bird, even made the inclusion of Chinese characters as a prerequisite function in their phones in response to the need of the local market (Hu 2000).

2005–2010 – the high prevalence of mobile phone usage and the diversification of the mobile phone market

By the end of 2010, the number of mobile phone users reached 85,900 million and the mobile penetration rate hit 64.6 per cent. Compared with 39,342.8 million in 2005, the figure had increased more than twofold (State Statistical Bureau 2010). With the increasing demand for mobile phones, more and more new companies strived to capture market share from the world's biggest mobile phone market. Some of them even tried to evade taxes, regulatory fees and safety checks and produced counterfeits and no-name knockoffs (also known as '*shanzhai*' or 'mountain fastness') so as to maximize and generate greater revenue with a built-in cost advantage (Shen 2008). An underground mobile market was gradually formed, which simply copied or mimicked brand names and their product designs such as Hiphone/Sophone for iPhone, Nokir for Nokia, Sonyericcson for Sony Ericsson, and Sumsung for Samsung (Barboza 2009). According to a survey conducted by All View Consulting, the number of mobile phones produced in China reached 6 billion in 2009; of these, 1.45 billion were knockoff mobiles (Xinhuanet 2010). In 2010, Nokia inevitably included knockoff mobile phones in their market report, stating that their market share was lowered from 38 per cent to 34 per cent because of the challenge from the counterfeit goods (Chinanews 2010).

Knockoff mobile phones are attractive not only because they are sold at a cheap price with low manufacturing cost, but also because they are able to fulfil the local market needs with eye-catching features. For example, many knockoff mobiles allow two subscriber identity module (SIM) cards to be simultaneously inserted in one machine, of which mobile users can choose different standby modes whenever and wherever they need. This specific function is very convenient and economical to those who need to travel between different provinces due to the implementation of the national roaming system in China, like migrant workers who may cross provinces to find jobs or return home to celebrate Lunar New Year and other special occasions; they can save more with dual-SIM mobile phones (Cheng in press). However, although knockoff phones have their own advantages, it does not necessarily guarantee that the grey handset industry can generate greater market share in the local market. According to a survey conducted by iSuppli, the number of knockoff phones produced in China reached 1.75 billion, while only 25 million were sold locally. The rest were mainly exported to India, South-east Asia, South America and Africa (Cctime 2010).

This reflects a situation that the competition on China's local mobile market has become keener. Although different strategies were employed to compete for a place in the market by enhancing competitive power such as chasing for the lowest price, merging with another company and cooperating with mobile service providers to form a selling alliance (Lu *et al.* 2007), some were finally eliminated through competition. For example, *Xiaolingtong*, which was formally known as PHS (Personal Handyphone System), adopted the micro-cellular technique of connecting to the network of fixed lines by wireless technology. In May 2000, the number of *Xiaolingtong* users increased rapidly when the MII announced that it was to suspend the business in every locality in order to protect China Mobile and China Unicom. In June 2000, the MII promulgated a 'Notice on Regulating PHS Development and Operation', which allowed the PHS as a 'supplement to and extension of fixed-line telephony' and restricted *Xiaolingtong* to be 'developed only in medium cities and small cities'. Although *Xiaolingtong* could only focus on the lower end of the market, including small towns and villages (Chen 2002), its emergence did create immense competition for China Mobile and China Unicom because of the huge potential of the mobile in China. In 2003, the growth rate of *Xiaolingtong* users surpassed that of both China Mobile and China Unicom. By the end of 2004, the number of *Xiaolingtong* users had reached 50 million. In 2006, *Xiaolingtong* found its peak of 1 billion users (Qiu 2007). In 2008, however, because of the increasingly intense competition and high maintenance and operation costs, *Xiaolingtong* users dropped to around 70 million, and by the end of 2010, it had only 30 million left (MC21st.com 2011).

Undoubtedly, cheap price and domesticated functions were the key reasons for the success of knockoff phones in the past few years; on the other hand, these two reasons contributed to the deficiency in quality, both in terms of technology and of production. Since knockoff phones were focused on employing a low-price strategy as a means of driving sales, doing so might limit their ability to innovate and expand their influence to other parts of the mobile phone market. The side-effect for the prevalence of knockoff phones is that China's local mobile market was not only threatened by the diminishing marginal profit because of market substitution, but also it faced a situation that bad products drove out the good. Especially when 3G mobile phones were launched onto the market, the lower competitiveness of local mobile phones was obviously reflected.

Before the birth of GRPS (General Packet Radio Service, also known as 2.5G), China's mobile users usually used WAP (Wireless Application Protocol) technology, which was set up in the WAP Forum in 1997 to enable users to transmit information between the Internet and a mobile phone to surf the Internet. On 25 March 2000, China Mobile officially began operating the commercial trial network for WAP in Shanghai, Beijing, Tianjin, Guangzhou, Hangzhou and Shenzhen. At the same time Guangdong Unicom developed their WAP business, for example in Beijing, Tianjin, Shanghai and Chongqing (Deng 2000). On 16 March, the first Chinese information service website (www.cwap.com) which could be browsed directly by a WAP mobile phone was established; more than ten ICPs (Internet Content Providers), including Sina, NetEase, eBay and others, announced their support for WAP website services (Deng 2000), while the Internet operators provided functions for downloading data such as stock information, weather forecasts, wallpapers and ringtones. However, since the length of time that it takes to upload and download information via WAP had restricted its further development (Chen 2000), by the end of 2000 the number of WAP users was just 50,000–60,000, about 2 per cent of the mobile phone market (Li 2000).

Until 2009, after the 3G licences were issued, the number of mobile netizens in China had increased significantly. In preparation for the adoption of the 3G technology, the Chinese government first tried to restructure the institutional formation of the mobile phone market in order to guarantee that no mobile company could monopolize the market and to ensure continuous technological advancement. In 2008, China Unicom, the country's second-largest mobile service provider, acquired fixed-line operator China Netcom Group Corporation with a share-swap deal. Since China Unicom originally operated both mobile and fixed-line services, the merger contributed to the new company's enjoyment of greater economic scale and ability to focus on developing its GSM technologies for the 3G services (Xinhuanet 2008). Later on, China Satcom was divided into two parts which merged into China Telecom. China's telecom market was restructured into three groups in order to pave the way for building new networks to promote expansion, and stimulate domestic consumption. Each group was then awarded 3G mobile phone licences: China's biggest mobile operator, China Mobile, for TD-SCDMA, the domestically developed 3G standard; China Telecom for the US-developed CDMA2000; and China Unicom for the European Wideband Code Division Multiple Access (WCDMA). The MIIT also issued 23 regulations on the requirements of 3G network operation, covering aspects such as market competition, consumer rights, user information security, telecommunication charges management and facility building (China Daily 2009). At the end of 2009, the investment in 3G infrastructure amounted to 1,609 billion RMB, which was greater than half of the size of investment accumulated in the past ten years (MIIT 2009). In July 2010, the number of 3G users had increased by a month to month growth of 11.4 per cent and increased 544.6 per cent compared with the same period the previous year, up to 280 million in total (MIIT 2010).

Summary

China was not the first country to market the mobile phone. However, it is the country where the development of the mobile phone has been the fastest and the most energetic. This chapter reviewed the development of mobile phone use in China from 1987, when the first mobile phone appeared, to the time of writing (2010). It describes the four-stage process of development against the background of macro-factors, such as the improvement of living conditions in China, which, it is contended, was one of the reasons for these improvements.

There is an old saying in China which advises, 'First know "what", and then know "why".' Therefore, the chapter started by providing information about the development process as broadly as possible from a wide variety of sources to describe the reality of the use of mobile phones in China today. Recently, more and more foreign scholars have joined the study of mobile phone development in China, not only because there is a huge market in China, but also because there are many interesting areas of study, such as why the market for mobile phones in particular flourishes more than other digital products in China, why the mobile phone industry continues to grow better in China than other countries, and how mobile technology may change or even transform Chinese society. What the future holds as far as the development of the mobile phone industry is concerned is unknown at this stage and predictions cannot be made without further data. However, through our field work on mobile phone use in China conducted by the Hong Kong Polytechnic University and Peking University, we have already found that the mainstream users (e.g. the middle class) consider the mobile phone as commonplace, whereas people who are not the mainstream members of society, such as migrant workers and white-collars, view mobile phones as essential to their wellbeing. Therefore, it is likely that the mobile phone market will continue to grow in the years ahead with a certain amount of stratification between different socioeconomic groups. Our future research will strive to uncover more detailed data in order to unearth these trends and the reasons for them.

References

Barboza, D. (2009) *In China, knockoff cellphones are a hit.* Online. Available: www.nytimes.com/2009/04/28/technology/28cell.html (accessed 28 February 2011).

Bi, W. (2000) 'Dilirenhe, zhenxingkeqi' [Favourable terrain and friendly people, promotion can be expected], *Xin Kuaibao* [New Bulletin], 14 July. Online. Available: www.ycwb.com/history/gb/2000/07/14/xkb/jjxw/5.html (accessed 25 February 2011).

Cctime (2010) *iSuppli: Shanzhai shouji 6 cheng xiaowang haiwai* [iSuppli: Sixty per cent of knockoff mobile phones were exported]. Online. Available: www.cctime.com/html/2010-11-10/20101110853537323.htm (accessed 25 February 2011).

Chen, D. (2002) 'Xiaolingtong weixie yidong tongxin' [*Xiaolingtong* is threatening mobile communications], *Tongxin Shijie* [Communications World], 11: 26–27.

Chen, R. (2000) 'WAP de jishu yu yingyong' [The technique and application of WAP], *Guangdong Zidonghua yu Xinxi Gongcheng* [Guangdong Automation and Information Project], 3: 34–37.

Cheng, C.T. (in press) 'Floating workers and mobile QQ: the struggle in the search for root', in L. Fortunati, R. Pertierra and J. Vincent (eds) *Migrations, Diaspora, and Information Technology in Global Societies*, London: Routledge.

China Daily (2009) *China's telecoms sector gets 3G licenses.* Online. Available: www.chinadaily.com.cn/business/2009-01/07/content_7375721.htm (accessed 28 February 2009).

Chinanews (2010) *1.2 yitai shanzhaiji ladi nuojiya shichang fene* [Nokia's market share was eliminated by 1.2 knockoff phones produced]. Online. Available: www.chinanews.com/it/news/2010/03-17/2173517.shtml (accessed 28 April 2010).

Deng, L. (2000) 'Zhongguo li WAP you duoyuan' [How far is between China and WAP?], *Zhengquan Shibao* [Stock Times]. Online. Available: http://webcache.googleusercontent.com/search?q=cache:o5K-EC0Fvb4J:past.people.com.cn/bike/viewnews.btml%3Fid%3D102923+%E4%B8%AD%E5%9C%8B%E9%9B%A2WAP%E6%9C%89%E5%A4%9A%E9%81%A0&cd=1&hl=zh-TW&ct=clnk&gl=hk&source=www.google.com.hk (accessed 5 June 2005).

Dong, D. and Song, L. (2000) *Zhongguo yidong tongxin hangye* [The mobile communications industry in China]. Online. Available: http://bi.icxo.com/htmlnews/2005/05/26/602308.htm (accessed 6 July 2008).

Hu, D. (2000) 'Shouji wenzi shuru ruanjian shichang ji fazhan' [The market of character input software and its development], *Tongxin Shijie* [Communications World], S1.

—— (2003) 'Guoji youshi lun' [Advantages of the domestic mobile phone], *Shuzi Tongxun* [Digital Communications], 19: 89–91.

Information Center of China Network (2003) *Shida hulianwanglu danwei jianjie* [A brief introduction of the top ten network companies]. Online. Available: www.cnnic.net.cn/html/Dir/2003/11/17/1330.htm (accessed 4 March 2005).

Jisuanji Shijie [Computer World] (2000) *Zhongguo yidong dianhua shichang jingzheng zhuangkuang fenxi* [Analysis of the competition situation on the mobile phone market in China]. Online. Available: www2.ccw.com.cn/2000/0006/0006e05.asp (accessed 4 March 2005).

Kuang, W. (2003) 'Lun shouji meiti' [On mobile phone media], *Guoji Xinwenjie* [The International Press], 3: 55–58.

Li, W. (2000) 'Qiantan WAP shouji de fazhan qianjing' [A brief talk on the prospect of WAP mobile phones], *Tongxin Shijie* [Communications World], 24: 38.

Lu, D. and Wong, C.K. (2003) *China's Telecommunications Market: entering a new competitive age*, Cheltenham: Edward Elgar.

Lu, Y., Dong, Y. and Wang, B. (2007) 'The mobile business value chain in China: a case study', paper presented at the International Conference on the Management of Mobile Business, Toronto, Canada. Online. Available: http://ieeexplore.ieee.org/xpls/abs_all.jsp?arnumber=4278568&tag=1 (accessed 16 July 2010).

MC21st.com (2011) *Zhongguo xiaolingtong yonghu fazhan qingkuang tubiao* [The development of *Xiaolingtong* users in China]. Online. Available: www.mc21st.com/~p.aspx?id=19 (accessed 6 July 2011).

MIIT [Ministry of Information Industry Technology] (1992, 1994, 1997, 1999, 2000, 2001, 2003) *Tongxinye Nianbao* [Telecom Industry Yearbook], China Statistics Press. Online. Available: www.miit.gov.cn/n11293472/n11293832/n11294132/n12858447/index.html (accessed 25 February 2010).

—— (2009b) *2009 nian woguo 3G he TD fazhan zongti qingkuang* [The development of 3G and TD in China in 2009]. Online. Available: www.miit.gov.cn/n11293472/n11293832/n11294132/n12858447/12979622.html (accessed 2 February 2010).

—— (2010) *2010 nian 7 yue woguo 3G yonghu he touzi pingwen zengchang* [The number of 3G users and investment increased steadily in July, 2010]. Online. Available: www.miit.gov.cn/n11293472/n11293832/n11294132/n12858447/13369489.html (accessed 28 November 2010).

Peng, C. (2005) 'Paizhao hezhunzhi wudong daqi, xiangguan xize yijing chutai' [The 'licence' system is waving its banner, and the concerning rules are coming], *Nanfang Dushi Bao* [Southern Metropolis Newspaper], 22 February. Online. Available: http://biz.163.com/05/0222/12/1D6O2HFU00020QFC.html (accessed 25 July 2005).

People (2001) *Zhongguo shouji yonghushu chaoguo riben, jiciyu meiguo* [The number of mobile phone users in China surpassed Japan's, only behind US's]. Online. Available: http://japan.people.com.cn/2001/03/06/riben20010306_2641.html (accessed 28 February 2005).

Qian, Q. and Yang, F. (2004) 'Shouji: anzhan, lianzhan, luanzhan' [Mobile phones: secret, protracted, tangled war], *Zhongguo Xinshidai* [China's New Era], 3: 78–83.

Qiu, L. (2007) 'The accidental accomplishment of Little Smart: understanding the emergence of a working-class ICT', *International Journal of Communication*, 1: 74–91.

Shen, S. (2008) *China's amazing knockoffs*. Online. Available: www.upiasia.com/Society_Culture/2008/12/09/chinas_amazing_knockoffs/7656/ (accessed 20 February 2009).

State Statistical Bureau (1988, 1993, 1995, 2002, 2004, 2010) *Zhongguo Tongji Nianjian (China Statistical Yearbook)*, China Statistics Press.

Wang, X. (1994) *Mobile Communication in China: an analysis of the diffusion process*, Ann Arbor, MI: UMI.

Wu, J. (2005) 'Zhongguo diyyiwei shouji yonghu hua dangnian' [The first mobile phone user talked about that year], *Xinxi Shibao* [Information Times], 26 April.

Xiang, L. (2001) 'Guochan shouji fazhan de wenti ji jiejue banfa' [Problems and countermeasures of the development of the domestic mobile phone], *Tongxun Shijie* [Communications World], 6: 4.

Xinhuanet (2008) *China Unicom unveils details of merger with China Netcom*. Online. Available: www.chinadaily.com.cn/china/2008-06/02/content_6729822.htm (accessed 8 February 2010).

—— (2010) *2009 nian zhongguo zao shouji 6 yi bu shanzhaiji chuhuo 1.4 Yi* [In 2009, China produced 6 billion mobile phones; 1.4 billion of them were knockoff phones]. Online. Available: http://news.xinhuanet.com/it/2010-01/06/content_12762979.htm (accessed 28 February 2011).

Xu, S. (2005) 'Paizhao hezhuizhi dui guonei shouji chanye yingxiang jihe' [What is the consequence of the 'licence' regulation to the domestic mobile phone industry?], *Renmin Youdian Bao* [People Post and Telecommunications Newspaper], 16 February. Online. Available: www.yesky.com/NetCom/218450974501502976/20050204/1909531_1.shtml (accessed 6 July 2006).

Yao Guozhang (1999) 'Zhongguo shouji shichang de xingcheng yu fazhan' [The shaping and development of the mobile phone market in China], *Dangdai Tongxin* [Modern Communication], 21: 4–15.

Yi, J. (2005) *Shouji de zuotian, jintian he mingtian* [The yesterday, today and tomorrow of mobile phones]. Online. Available: www.yesky.com/Fashion/73747556294721536/20050406/1932077_6.shtml (accessed 25 February 2009).

Yidongju [Mobile Bureau] (1998) 'Zonghe baodao: Zhongguo GSM quanqiutong shuzi yidong dianhua jiang yu 1999 nian shengwei' [Comprehensive dispatch: GSM mobile phones in China will increase in 1999], *Xiandai Dianxin Keji* [Modern Telecommunications Technology], 10: 45.

Yin, J., Tang, L. and Yin, Y. (2004) 'Diwu meiti: shehuixue shijiao de fenxi' [The fifth media: analysis from a sociology perspective], *Xinan Minzu Daxue Xuebao-Renwen Sheke Ban* [Southwest National University Journal-Edition of Arts and Social Science], 3: 320–21.

Yu, S., Wu, T. and Ge, Z. (2001) *Dianhua chuzhuangfei he shouji ruwangfei youhui tiaozheng* [The preferential adjustment of the instalment fee of fixed line and network access fee of mobile phones]. Online. Available: www.chinanews.com.cn/2001-07-02/26/101845.html (accessed 7 August 2005).

Zhang, Y. (2001) 'Guochan shouji qiye shounao luntan chengli' [The forum of head of domestic mobile enterprises established], *Guangming Ribao* [Guangming Daily], 2 June.

Zhu, L. and Chen, B. (1999) 'Zhongguo shouji shichang jingzheng taishi' [The competition situation on the mobile phone market in China], *Xiaoshou Yu Shichang* [Sale and Marketing]. Online. Available: www.emkt.com.cn/article/2/275.html (accessed 25 February 2005).

6 Comparative development of the mobile Internet in China and Japan

Eric Harwit

Introduction

Over the past 15 years, mobile phone use has risen dramatically in China, and the country today has the largest mobile phone population in the world. But the PRC is now entering a new level of mobile phone use, as third-generation (3G) mobile phones offer access to high-speed Internet information. How will mobile phone Internet access shape usage patterns once the new technology becomes common?

To better understand the ways the mobile Internet network may grow in China, this chapter will first explore China's East Asian neighbour, Japan, to examine how the Japanese have adopted and utilized mobile phone Internet technologies. The Japanese system, pioneered by NTT DoCoMo and other corporations, has transformed many aspects of Japanese social, cultural, and commercial life. The chapter will consider the ways Japanese use the mobile Internet, on a small screen, for basic communication, group interaction, and mobile commerce, and examine the successes and failures of the Japanese mobile Internet since its introduction at the beginning of the decade.

The chapter will then examine the development of China's mobile Internet in light of the Japanese experience. Do the early stages of mobile phone Internet use in China mirror those of Japan? What similarities and differences can we expect as mobile Internet use further penetrates the Chinese population? In particular, what lifestyle differences and consumer preferences will be the key in shaping the way Chinese citizens adopt 3G mobile phone data communication? Will there be different trajectories in urban areas of China (built in a pattern geographically similar to that of Japan) versus the less concentrated and less affluent rural parts of the nation?

The more restrictive political system in China will also limit the ways the mobile Internet can be used in the PRC, and the chapter will consider how these restrictions will shape 3G use in the early years of the technology. But a lively arena already exists via personal computers (and, in nascent form, mobile phones) for discussion of cultural, social, economic, and non-sensitive political issues that affect many aspects of Chinese life. As China's political and social liberalization continues on a steady course, we may see some convergence in the ways mobile telephone Internet use grows in both China and Japan.

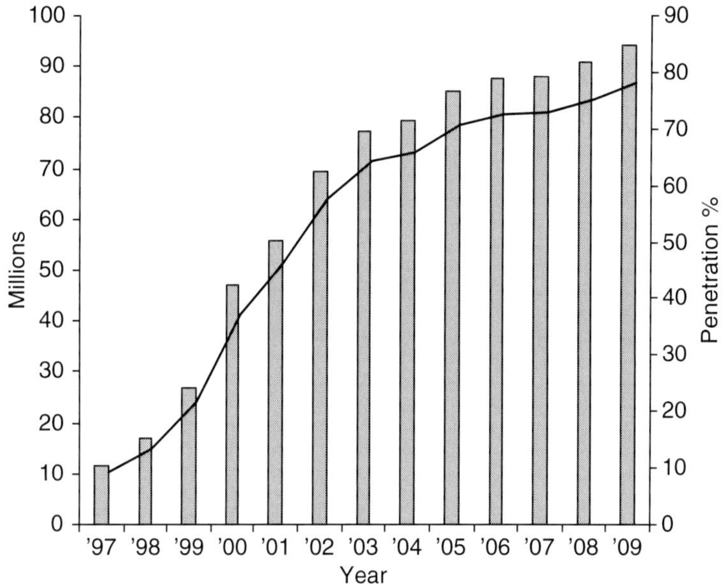

Figure 6.1 Internet users in Japan (1997–2009). Bar is number of users; line is penetration percentage.

Source: Ministry of Internal Affairs and Communications (2010: 41).

The Internet and mobile Internet in Japan in the past decade

Before turning to the ways the mobile Internet grew in Japan, we first consider briefly the expansion of the more traditional fixed-line data network service in the country.

From 1995, Japan had already started to lag behind other nations in Internet development. In March of that year, all of Japan had only about 300 www sites, while the United States had about 9,000 (Coates and Holroyd 2003: 27). In 1997, only about 12 million people, or 9 per cent of the Japanese population, used the Internet, while about 20 per cent of Americans and northern Europeans had become 'netizens' (Ministry of Public Management, Home Affairs, and Posts and Telecommunications 2004: 12; nua.com Survey A, Survey B). Figure 6.1 charts the growth in overall numbers of Internet users in Japan since the late 1990s.

Why were the Japanese relatively slow to adopt home computer-based Internet in the 1990s? First, prices for access on home computers were high. NTT, the telecom monopoly until recent years, charged a standard 10 yen (about 10 US cents) for 3-minute local calls (Kyodo News Service 1996), and up to the late 1990s, monthly Internet service provider fees for dial-up access started at 4,000

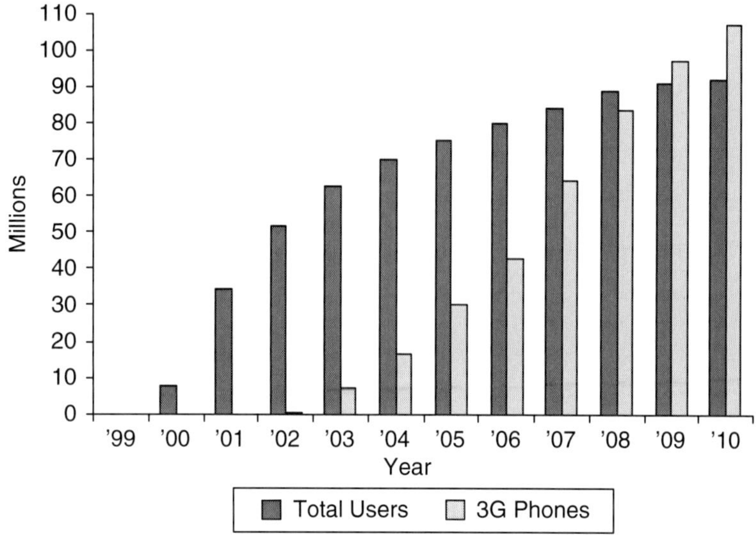

Figure 6.2 Mobile Internet users and number of 3G phones in Japan (1999–2010). The number of 3G phone subscriptions may be greater than mobile Internet users, as some users have more than one 3G subscription.

Sources: For 1999–2005, Ministry of Internal Affairs and Communications (2005: 34); for 2006–2009, Japan Statistical Yearbook (2011: section 11–6); for 2006–2010 3G phones, Business Monitor International Communications, various reports; for 2010 total users, Budde.com report (2010). (Statistics are generally consistent between sources.)

yen ($35) per month. Modest users paid as much as 15,000 yen ($130) per month for slow modem access (The Economist 1999).

Other reasons included the Japanese lifestyle. Office workers often spend long hours commuting from home, and leave home early and arrive back late. Even children have relatively little leisure time at home, as many attend *juku* (after-hours schools) on afternoons and weekends. Also, many Japanese were not accustomed to typing on a computer keyboard.

The advent of NTT's DoCoMo mobile phone-based Internet service, called 'i-mode', served to jump-start the use of the Internet in Japan at the turn of the decade. The service began in February 1999, and by December 2000 had more than 14 million subscribers. This represented about one-third of all Internet users in Japan at the time. Figure 6.2 indicates the rapid growth of DoCoMo and other mobile Internet company customer numbers.

What was the attraction of the new service? First, it was relatively cheap. The i-mode monthly fee in 2000 was only 300 yen per month (about $3), and an additional use fee was based on the amount of information downloaded, charged in fractions of yen per data packet. A brief e-mail would cost only about 1 yen. And the phone itself cost a maximum of 20,000 yen ($190) (Kageyama 2000).

The mechanics of i-mode also made it more accessible to those unaccustomed to using the Internet. I-mode service, unlike modems, was always on – subscribers needed only to touch the button that said ‘i’, and they could begin receiving or sending data. Japanese kanji and hiragana were relatively easy to input for short messages, and the compact nature of the characters (they do not have to be scrolled around on a small screen, as English does) makes them more amenable to a mobile phone. This also applied to Chinese characters, of course, and has likely had implications for Chinese mobile phone use, which is discussed later in this chapter.

DoCoMo was also careful to offer useful content. The company provided many in-house websites that were tailored to customer needs, and had many traditional www sites configured for the small screen. By 2002, the company offered more than 3,000 i-mode sites, and access to 60,000 other unregulated Internet web pages (Coates and Holroyd 2003: 74).

One significant difference in mobile user profiles in Japan was the rapid adoption of the service by older customers. The number of users aged 20–24 was almost equal to those aged over 40 (ibid.). It is likely the ease of use contributed to this phenomenon.

Japan’s rivals in fixed-line service provision saw the great success of DoCoMo’s mobile service, and finally moved to offer competing access to those who preferred to use home computers. The breakup of NTT in 1999 contributed to the rise of rival corporations providing telecommunication services to home users. In 2001, several companies began to offer a high-speed Asymmetric Digital Subscriber Line (ADSL) Internet service to home PC users. Yahoo Japan, NTT East and NTT West, and other companies launched a price war to attract customers, taking monthly fees down as low as 2,000 yen ($19) per month (Nikkei Weekly 2001). In 2002, the television cable company Jupiter Telecommunications (also called J-COM) offered telephone plus broadband Internet access for about 2,700 yen ($26) per month (Multichannel News International 2002: 20).

By 2003, the combination of mobile, home, office, and school Internet users put Japan among the ranks of most other developed nations for penetration of data network use. But as Figures 6.1 and 6.2 indicate, in that year fewer than 10 million of Japan’s total 77 million Internet users accessed the network solely on a PC, as the mobile phone Internet population neared 70 million. Most connected to the data network from both mobile phone and PC, but in 2003 about 7 million people (or about 10 per cent of netizens) only used their mobile phone for Internet access (White Paper on the Internet in Japan 2002). In 2011, a survey found a continuing preference of Internet access via the PC, as some 75 per cent of users said they used a computer for web browsing, while only 15 per cent said they mainly used a mobile phone for this activity (Japan.Internet.com Survey 2011). By 2009, of 94 million Internet users, 91 million had mobile Internet access.

As one might expect, Japanese mobile Internet customers have had somewhat different use patterns from fixed PC users. Figure 6.3 shows how the mobile phone functions differed from those of PCs in 2003, and how mobile Internet functions have developed over the past decade. As the figure indicates, the predominance of e-mail usage on PCs has carried over into the mobile phone

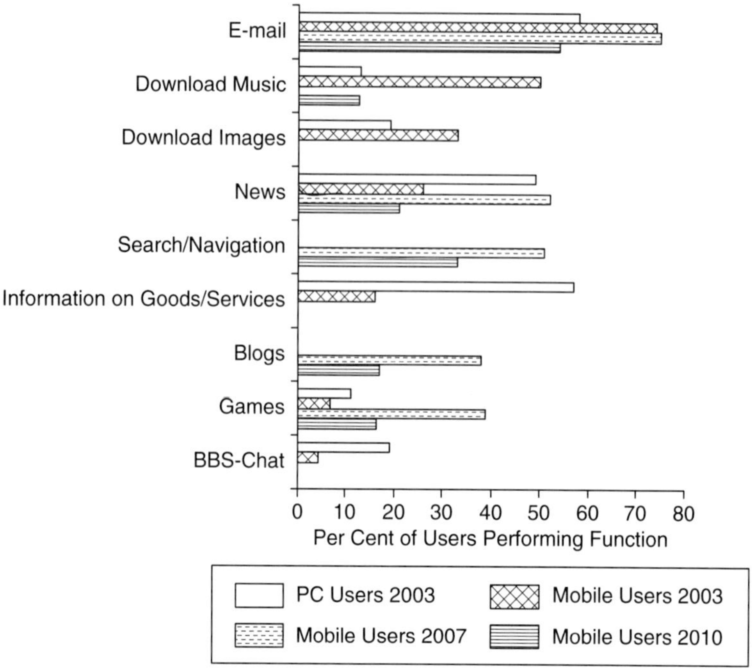

Figure 6.3 Trends in selected Internet use from mobile phones and PCs in Japan. Not all categories are available for each year.

Sources: For 2003 data, Ministry of Public Management, Home Affairs, and Posts and Telecommunications (2004: 12–13); for 2007 data, comScore press release (2007); for 2010 data, comScore press release (2010).

domain. The rate of mobile users downloading images and using the Internet to play games is similar for both mobile and fixed Internet users.

The main differences that emerge are related to the physical and spatial characteristics of the two network access methods. Users in 2003 were more likely to employ their portable phones to download or listen to music. This is based on the user lifestyle: long commutes by train or subway mean the mobile phone was a convenient tool for entertainment; at home, users have alternative devices to listen to music. By 2010, however, commuters could likely use alternative music sources without having to download music on their phones, and the use of downloading music via a mobile handset fell. For citizens on the go, use of the mobile Internet for gathering news and for search/navigation functions continued to be popular options.

The commuting lifestyle is also reflected in the use of mobile phone e-mail: while about 87 per cent of Japanese mobile Internet users in 2005 sent e-mail, only about 12 per cent of American mobile phone users sent e-mail. Even in South Korea, which also had a high percentage of advanced-generation mobile phone users, only about 43 per cent of users sent e-mail on their phones (Ministry of Internal Affairs and Communications 2005: 17).

For searches of information on goods and services, however, the home or office environment, with a larger PC screen, provides a more appropriate environment for finding and perhaps comparing required data. The low rate of Bulletin Board System (BBS) and blog use on mobile phones also indicates the difficulty or inconvenience of using this tool on a mobile device. As the chapter will discuss later, Japanese use of BBS or chat services lags behind that of Chinese network users, and the mobile penetration of these services is even lower in Japan.

The state of Japanese mobile use, 2005–2010

The trend toward increased use of mobile phones for Internet access in Japan has shown steady growth in recent years. At the beginning of 2005, DoCoMo had about 48 million subscribers, and rival companies with similar services had another 40 million or so. About 86 per cent of all Japanese mobile phone users subscribed also to a mobile Internet service (ibid.: 34).

A major change in the way mobile Internet services were billed came about in late 2003, as KDDI introduced fixed-rate packet services for its mobile Internet. No longer did customers have to worry about large fees for downloading significant amounts of data. KDDI charged 4,200 yen (or $39) per month for unlimited use (Nikkei Weekly 2004). DoCoMo responded to KDDI's challenge in June 2004, by offering the same service for 3,900 yen ($36) per month (Japan Corporate News Network 2004). By October 2004 there were about 3 million subscribers to the fixed-packet service (JP Morgan 2004: 6).

As of early 2005, about 9 million Japanese used DoCoMo's faster 3G mobile phone service – this was about 18 per cent of the company's 48 million subscribers. The average user paid about 10,000 yen per month ($95) for the 3G service, and 7,000 yen ($65) for 2G (Nikkei Weekly 2005a). The 3G service offered about 400 kbps download speed, but newer versions increased this to more than 1 Mbps. DoCoMo's future 4G network promises download speeds of 100 Mbps to 1 Gbps (Financial Times Information 2004), and by 2010 its LTE (Long Term Evolution) service had already reached nearly 40 Mbps (NTT DoCoMo 2010). In 2008, DoCoMo, KDDI, and Softbank Mobile all stopped taking new 2G subscribers. For DoCoMo, 3G users topped 45 million, and 2G subscriptions had shrunk to about 8 million (BMI 2008: 1 October). As Figure 6.2 indicates, by 2010 Japan hosted about 110 million 3G mobile phone subscriber contracts.

As we will see later, China was until recently delayed in moving to 3G by a debate over which transmission standard to employ. In Japan, however, two variants of the globally dominant CDMA (Code Division Multiple Access) standard grew rapidly. As of early 2005, CDMA 2000 (first offered in April 2002) was used by 17.93 million 3G customers, and W-CDMA (launched in October 2001) by 12.42 million (Ministry of Internal Affairs and Communications 2005: 34; Ministry of Public Management, Home Affairs, and Posts and Telecommunications 2004: 3). In 2010, the division remained, with 77 million users of W-CDMA, and 32 million CDMA-2000 (Telecommunications Carriers Association of Japan Telecom Data Book 2010: 10).

DoCoMo rival KDDI initially had a far higher penetration of 3G users, with about 90 per cent of its customers using the newer generation of its 'au' brand service. The company's main selling point was its popular offer of song downloads (Nikkei Weekly 2005a). Overall, however, in the middle of the decade KDDI trailed DoCoMo as Japan's second largest mobile phone company, with 19.54 million subscribers; Vodaphone (15 million) and TU-KA (3.58 million) trailed (Jiji Press 2005). By 2010, DoCoMo had nearly a 50 per cent market share, with KDDI at 28 per cent, Softbank Mobile (which had bought Vodaphone Japan in 2006) at 20 per cent, and EMOBILE at only about two per cent (Telecommunications Carriers Association of Japan Telecom Data Book 2010: 10).

Despite the move to a new generation of mobile Internet service, Japan is beginning to see market saturation. Already by 2005, some two out of three people (or 87 million) owned mobile phones (the number rose to about 95 per cent of the population in 2010), and handset makers such as NEC were predicting losses in phone sales for 2005 and into future years (Jiji Press 2005; Nikkei Weekly 2005a).

Which services stand to be profitable in Japan's mobile Internet era? As in many countries, Japanese network users are reluctant to pay for content or services that they might be able to find at no cost. Nearly 75 per cent of PC users reported in 2005 they had never paid for Internet information, though only about 56 per cent of mobile Internet users had the same response (Ministry of Internal Affairs and Communications 2005: 23).

For mobile phone users in 2005, the most popular feature subscribers were willing to pay for was downloading ring-tone melodies. About 36 per cent of Japanese mobile Internet users paid to personalize the way their phone sounds. The advent of 'ring-back' tones, in which the person calling the subscriber hears a song or sometimes a joke rather than the traditional ringing sound, promised to further add to the market. Ring-back tones only debuted in 2004 in Japan, after some 10 million South Korean subscribers led the market; some 500,000 Japanese had signed up for the service within the first two months of its availability (JCN Newswire report 2004).

On-line mobile games present the other major feature mobile phone Internet customers are willing to pay for. Some 15 per cent of users had paid to download games as of 2005. Games are known as 'sticky' features, as users tend to return to them over time, in particular if winning a game requires repeated attempts or if they become part of a team effort with friends or colleagues (Coates and Holroyd 2003: 71–72). Some fixed-station games have been adopted for mobile Internet use: DoCoMo, for example, announced in July 2005 that it would adopt the hugely popular 'Monster Farm' game to its 3G mobile phone network (Mainichi Daily News 2005: 8).

Other forms of mobile phone commerce will expand as new technologies become more widely available. For example, the Sony Corporation in 1999 introduced its 'FeliCa' electronic money cards, but not until 2004 did the technology move into the realm of mobile phone functions. As of August 2005, about 4 million 'wallet handsets' (or *osaifu keitai* in Japanese) were in use (Nikkei Weekly 2005c). By 2009, DoCoMo alone had nearly 35 million customers signed up for the service (plusd.itmedia.co.jp News Report), and rival companies offered a similar option.

Wallet handsets provide a new dimension for mobile phone use, as they can be employed to pay for some Japanese bus and railway tickets as well as at a growing number of retail outlets in Tokyo and other cities. They are also being integrated into Japan's ubiquitous network of vending machines ('Contactless' with Sony's FeliCa). In its present form, FeliCa is not clearly part of the mobile Internet data network, but cash transfers to accounts are already done through PCs, and it seems feasible that future functions would allow Internet capable phones to serve as tools to add value to a consumer's FeliCa-based account. Overall, Japan's mobile commerce market in 2009 reached 968 billion yen (about $10 billion), up from 171 billion yen ($1.4 billion) in 2003 (Ministry of Internal Affairs and Communications 2010: 45).

Despite the tremendous forward momentum of the mobile Internet, there are some factors that may slow its advance in the coming years. Already noted was the near saturation of mobile phones in Japanese society – even with new functions, users may be somewhat slower than in past years to purchase an expensive new handset to acquire additional features. The introduction of higher-speed networks may work to alleviate this problem, at least in the short term.

A further problem may come about if website content providers fail to configure their pages to the small screen. The popular search engine Google, and the on-line shopping mall Rakuten, for example, were not initially available to mobile Internet users (JP Morgan 2004: 9), though in 2006 Rakuten was included in DoCoMo's coverage, and by 2008 Google had forged partnerships with most of the leading Japanese mobile carriers. Moreover, according to one analyst, there were a growing number of popular mobile-only websites for which the content providers do not provide a traditional PC-accessible page (Interview 2005).

Finally, in a bridge to the Chinese case, we consider the assertion that the mobile Internet represents a 'poor man's PC' (JP Morgan 2004: 11–14). In Japan (as in many other countries, including China), a typical desktop or laptop computer costs a minimum of 40,000 yen (about $500). By contrast, a mobile phone in Japan may cost 40,000–50,000 yen, but carrier incentives normally cut that cost to about 20,000 yen ($180) or less. Some basic handsets, even ones that provide e-mail and other Internet functions, have been essentially free when the customer signs a user contract. In 2004, the penetration rate of mobile phones for people with incomes under 3 million yen ($27,000 at the time) was about 60 per cent, while only about 35 per cent of those same people owned PCs.

In Japan, then, many avid mobile Internet users seem to be those on low incomes, typically young people and high-school and college students, who have had their first Internet experience using a mobile phone. Use of the mobile

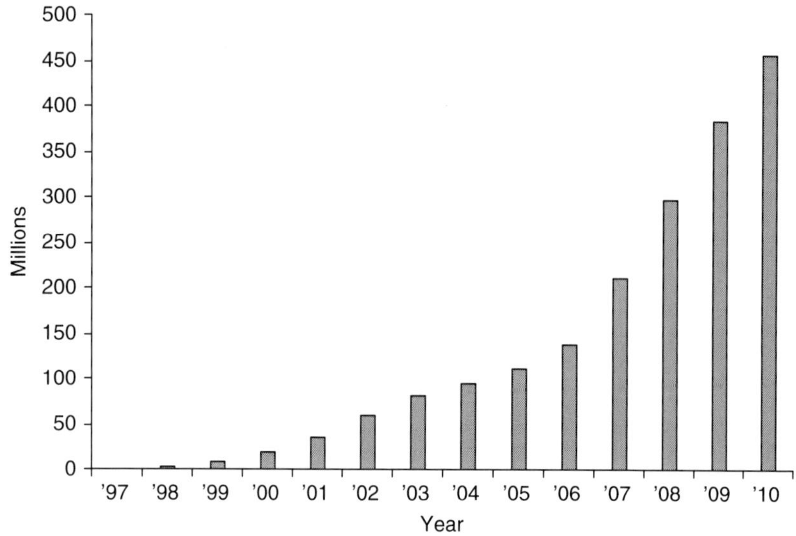

Figure 6.4 Total number of Internet users in China (1997–2010).

Source: China Internet Network Information Center, The People's Republic of China, various years.

Internet becomes integrated into their lifestyle. The JP Morgan study, building on this theme, forecasts five 'killer content' areas for those accessing the mobile Internet (some of which have already been highlighted in the earlier part of this chapter): ringtone melodies, games, custom screens, horoscopes, and dating services (JP Morgan 2004: 13). In the case of Japan, these extra services were often very inexpensive; in the case of ring tones, for example, the cost in 2004 was only 100 yen (90 cents) each (Japan Times 2004). Mobile phone comics (*manga* in Japanese) were also becoming popular by the middle of the decade, and cost users as little as 20–50 yen (18–45 cents) per story (Nikkei Weekly 2005b).

The mobile Internet in China

Unlike Japan, with its more than 10-year history of large-scale mobile Internet use, China is in the early stages of entering the realm of handset-based network use. The analysis here juxtaposes many of the developments in Japan, and projects them onto the China case. Which elements of Japan's trajectory will fit in for Japan's East Asian neighbour, and which will not?

Before turning to the issue of the mobile Internet in the PRC, we first give an overview of China's PC-based Internet growth over the past decade. Figure 6.4 shows the overall increase in users over the past 14 years. In 2005, China passed Japan to have the second largest number of Internet users in the world, after the United States. However, Figure 6.5 indicates that, as a percentage of the population, China still trails rather far behind Japan.

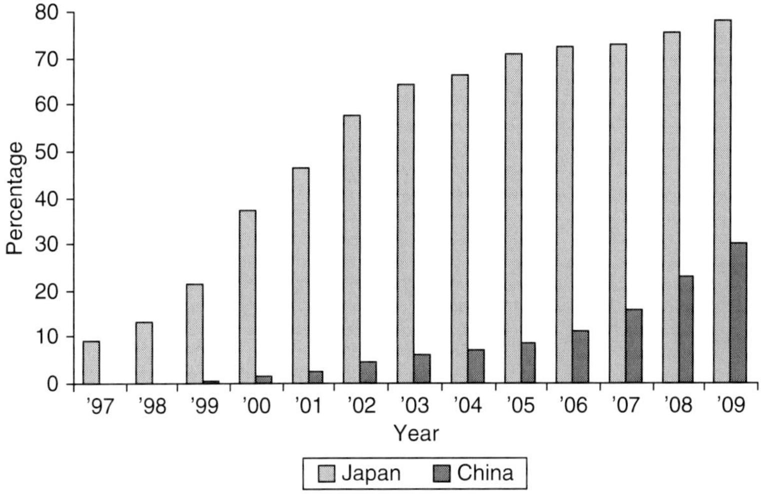

Figure 6.5 Internet penetration, Japan and China (1997–2009).

Sources: For Japan: Ministry of Internal Affairs and Communications (2010: 41); for China: China Internet Network Information Center, The People's Republic of China, various years.

As in Japan, the growth of Internet use in China was limited until the late 1990s by price (Harwit 2004; Harwit and Clark 2001). The cost of a dial-up service averaged about RMB400–600 ($50–75) for 40 hours of use per month in 1998, a high price when average incomes were less than RMB800 per month ($100) even in the wealthiest cities. In March 1999, Chinese premier Zhu Rongji ordered that several costs associated with Internet provision be lowered, with the result that monthly bills for an hour or two of modem use per day averaged about RMB160 ($20) per month. This was still relatively high, but did enable the rapid growth in users seen in Figure 6.4.

The proliferation of competitive Internet cafes, with user charges of as little as RMB1 per hour, as well as essentially free access for users in their workplace, meant the pool of Internet users could grow even faster. Internet use spread even more quickly as broadband access in Chinese homes made its debut in about 2001. As in Japan, the breakup of the monopoly China Telecom into two companies, China Netcom and the southern sector of China Telecom, spurred competition for broadband customers. Today, Chinese subscribers in large cities can get high-speed home Internet connections, mostly ADSL lines, for as little as RMB100 ($12) per month.

Figure 6.6 indicates how the Chinese use the Internet. In comparing the numbers with Figure 6.3, we see that Chinese PC use patterns are quite similar to those in the Japan case. For both countries' netizens, sending e-mail and getting news are vital functions. The Chinese, however, seem more likely to play on-line games or

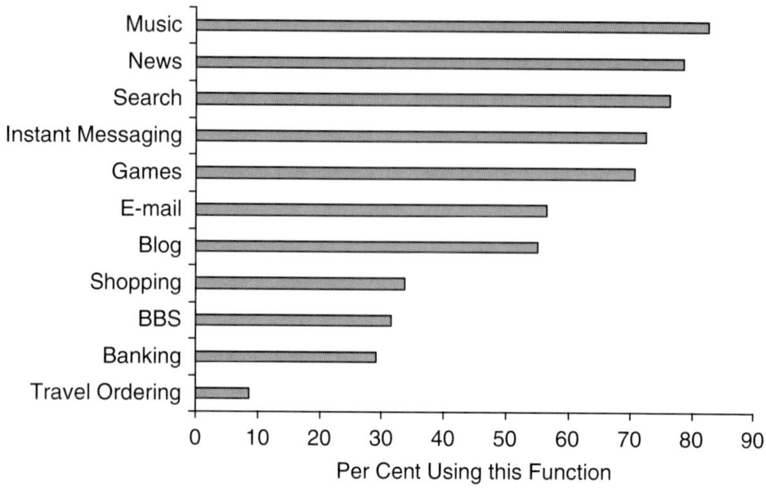

Figure 6.6 Selected Internet uses in China, mid-2010.

Source: China Internet Network Information Center, The People's Republic of China, July 2010 report.

use BBS functions than do the Japanese – this may be a function of young Japanese users having less time to enjoy the entertainment functions of the Internet.

More importantly for this study, we ask: How do Chinese access the Internet? As of June 2010 about 364 million Chinese used a broadband connection for Internet access, with the previously popular dial-up service shrinking. Some 88 per cent of users accessed the Internet at home, 33 per cent at their office or school, and about 34 per cent used Internet cafes (China Internet Network Information Center 2010). A more significant comparison in the current context is that, in mid-2010, about 66 per cent of Chinese Internet users (or nearly 280 million people) accessed the Internet via mobile phone (ibid.).

Though the number of mobile Internet users in China is now high, as recently as 2005 mobile Internet penetration was as low as 2 per cent (China Internet Network Information Center, July 2005 report). Why was mobile Internet access slow to take off in China? There was no shortage of mobile phone users; China even then led the world with more than 340 million mobile phones (Xinhua Economic News Service 2005b). There are several reasons for the slow move to mobile phone Internet use. First, China did not have the equivalent of 'i-mode'. There were, of course, methods to send short messages on Chinese mobile phones; in 2005 users sent about 20 billion messages per month (Xinhua Economic News Service 2005a). In fact, this use likely slowed the growth rate of Internet use in the middle of the past decade, as Figure 6.4 indicates. Short message sending technically does not count as Internet use, as the messages need not pass over the data network in order to be sent and received. In a sense, short messages substituted for either home or mobile Internet use; as in the Japanese case, sending e-mail (or the equivalent short message) is a top priority for Chinese communication.

Another reason there was relatively little mobile Internet use in China was the speed of the mobile network's data transmission. China (like the United States at the time) used 2G or '2.5G' mobile phone systems; this allowed limited access to websites, at very slow speeds. In 2005, the dominant mobile phone company in China, China Mobile, offered General Packet Radio Service (GPRS) transmission for its mobile phones, but the download speed was only 25 kbps. The only other licensed mobile phone company, China Unicom, had the faster CDMA1X service, with phones priced relatively cheaply at RMB588 ($73), but this gave the customer only about 60–70 kbps (Asia Pulse 2005; Financial Times Information 2005). As noted above, DoCoMo's 3G network offered Japanese customers 400 kbps download speed in 2005.

With few customers, Chinese website developers had little incentive to improve what mobile sites there were. The contrast with DoCoMo was sharp, in that the Japanese encouraged website creators to conform to the network that was already built up.

Despite the slow download speeds, by the middle of the decade Chinese mobile Internet customers were imitating some of the Japanese usage patterns. The company Kongzhong.com, for example, offered mobile phone users the popular game 'Duobao', and had 100,000 subscribers paying RMB5 (about 60 cents) per month as of March 2005. The total mobile phone on-line game market in China stood at about RMB600 million (about $74 million) in the first half of that year (SinoCast China IT Watch 2005a). Kongzhong's rival, Tom Online, sold $57.9 million in ring-back tone and other voice and 2.5G services in 2004, a giant leap from only $4.2 million sales for the same services in 2003. Tom Online was number one in China in 2005 for selling ring-back tones and interactive voice services (such as serialized stories), and trailed only Kongzhong for 2.5G services (Tom Online Inc. 2005: 7–8).

Change began to come to China's mobile Internet community in the latter part of the past decade. Only in 2009 did China's major phone companies begin to offer access to 3G networks, allowing mobile users the same kind of transmission speeds seen in Japan. The country was slowed by its need to choose between three competing standards, namely the European W-CDMA, American-developed CDMA2000, and China's own TD-SCDMA. The government was also in the midst of reorganizing the major mobile and fixed-line phone companies, and there was competition for licensing these companies as mobile carriers. Still, the higher speeds had facilitated an increase in mobile gaming: for the first half of 2010, mobile on-line game revenue reached RMB4 billion (about $600 million) (SinoCast Daily Business Beat 2010).

Despite the large number of mobile Internet users in the PRC, usage patterns are only now beginning to appear, and foreign companies may have some influence in the field. By 2005, DoCoMo had already established offices and investment in China, and awaited the advent of 3G to fully develop its presence in the PRC (SinoCast China IT Watch 2005b). China's three mobile phone companies – China Mobile, China Unicom, and China Telecom – began severe competition for 3G customers in 2009. Still, as of mid-2010, the relatively high cost of 3G access in China had limited its spread (China Internet Network Information Center 2010). As of February 2010, the country hosted only 16.1 million 3G users (Cheung 2010). One assessment in 2011 indicated that the Chinese may wait for a fourth-generation (4G) mobile technology to see a rapid spread of high-speed mobile Internet utilization (Xia 2011).

What will the effects be of an eventually expanded 3G or perhaps 4G mobile Internet use on China? As in Japan, the mobile Internet will likely bring a new wave of Internet users, ones who already have a mobile phone, and want to upgrade to mobile Internet access. As the cost for 3G phones and Internet access comes down, those Chinese users who cannot afford PCs and do not want to add broadband Internet fees to their monthly bills can almost instantaneously become Internet users. As early as June 2005, for example, 3G handsets made by the domestic Chinese company Huawei sold for as little as $99 in Singapore (SinoCast China IT Watch 2005c). Thus Chinese may follow the Japanese pattern of the mobile Internet being the 'poor man's PC'.

Mobile Internet use could also play a role in alleviating some of China's growing urban/rural network access division. For example, the phones could help bring Internet access to many places in China with low income levels but good public mobile phone systems, thus helping bridge the urban/rural 'digital divide'. After all, in mid-2010, only about 27 per cent of Chinese rural inhabitants were Internet users (China Internet Network Information Center 2010).

The Japanese example has shown that mobile users seem to grasp 'i-mode' use rather quickly, so there could be a shorter learning curve for those Chinese users not familiar with the more complicated PC Internet interface. In fact, those many millions of mobile phone users who now send text messages would likely find it to be only a short hop to using the mobile Internet. As in Japan, too, ease of use may encourage older Chinese to use the network; as of mid-2010, about nine per cent of Chinese Internet users were older than 40, though only 1.3 per cent were older than 50 (China Internet Network Information Center, July 2010 report).

Conclusion

The implications for broader mobile Internet access in China are not clear, but at a minimum we could expect trends similar to those in Japan, with older and less wealthy Chinese, ones who now are familiar with mobile phones, adopting the mobile Internet. As in Japan, we could also see more mobile users finding entertainment by downloading music, and, as connection speeds increase, using the Internet for BBS or chat participation.

It is too soon to predict exactly the ways the new 3G mobile Internet will affect Chinese society, but it is likely that the new technology will present both opportunities and challenges to the Chinese social fabric. Government censorship of PC-based Internet content has been widely reported in the Western media, and it is likely officials will extend their watch over politically challenging content into the mobile Internet sphere. Services such as chat rooms and mobile websites will likely face continued government scrutiny.

In the short run, however, the rather politically neutral features of music download, games, jokes, and other entertainment features will fall outside of government censorship hands. The adoption of commerce tools, such as FeliCa, will also probably be encouraged by a government that seeks to stimulate a consumer society.

Of course, there are many features of Japan and China that inspire a divergence of mobile Internet use. The conservative Chinese attitude toward adult-content or perhaps dating services may make the PRC system slow to emulate the Japanese services. Chinese workers are not now currently such a commuting-based force as one sees in major Japanese cities, and this could slow the demand for the kind of services Japanese use while sitting on a train or subway. However, most major Chinese cities are also building or expanding mass transit systems, so we may see a greater overlap in consumer demand between the two countries in the coming years.

Despite the vast differences between the Chinese and Japanese political systems, we may expect similar developments and trends in the two nations' mobile Internet systems. If DoCoMo or other Japanese carriers are allowed greater access to the Chinese market as the country adopts its high-speed 3G network, the similarities between the two countries' networks will likely be striking, and trends in Japan will form a roadmap for China's future mobile Internet development.

Acknowledgement

The author would like to thank Manca Sustarsic for research assistance with this chapter.

References

Asia Pulse (2005) 1 June report.

BMI [Business Monitor International Communications], various reports.

Budde.com report (2010) Online. Available: www.budde.com.au/Research/2010-Japan-Telecoms-Mobile-Broadband-and-Forecasts.html?r=51 (accessed 15 January 2011).

Cheung, J. (2010) 'The cost of connecting in China', *Forbes*, 4 April report. Online. Available: www.forbes.com/2010/04/14/china-mobile-3g-network-markets-equities-telecom.html (accessed 15 January 2011).

China Internet Network Information Center, various year reports. Online. Available: www.cnnic.cn/en/index/0O/02/index.htm (accessed 20 December 2010).

Coates, K. and Holroyd, C. (2003) *Japan and the Internet Revolution*, Houndmills, Basingstoke: Palgrave Macmillan.

comScore press release (2007) 'Mobile phone web users nearly equal PC based Internet users in Japan', 20 September. Online. Available: www.comscore.com/Press_Events/Press_Releases/2007/09/Japan_Mobile_Phone_Usage (accessed 15 June 2009).

—— (2010) 'comScore launches mobile measurement in Japan', 22 September. Online. Available: www.comscore.com/Press_Events/Press_Releases/2010/9/comScore_Launches_Mobile_Measurement_in_Japan (accessed 3 December 2010).

'Contactless' with Sony's FeliCa. Online. Available: www.sony.net/Products/felica/about/index.html (accessed 20 January 2010).

The Economist, 18 December 1999.

Financial Times Information (2004) 20 December report.

—— (2005) 22 February report.

Harwit, E. (2004) ‘Spreading telecommunications to developing areas in China’, *China Quarterly*, 180: 1010–30.

Harwit, E. and Clark, D. (2001) ‘Shaping the Internet in China’, *Asian Survey*, 41: 377–408.

Interview (2005) with telecommunications company analyst in Beijing, July.

Japan Corporate News Network (2004) 29 July report.

Japan.Internet.com survey (2011) Online. Available: http://japan.internet.com/research/20110311/1.html (accessed 25 March 2011).

Japan Statistical Yearbook (2011) *Japan Statistical Yearbook 2011*, Section 11–6. Online. Available: www.stat.go.jp/data/nenkan/zuhyou/y1106000.xls (accessed 3 March 2011).

Japan Times, 11 December 2004.

JCN Newswire report, 24 February 2004.

Jiji Press, 7 April 2005 report.

JP Morgan (2004) *Asia Pacific Equity Research (2004) Report ‘Mobile Internet Sector’*, 13 December.

Kageyama, Y. (2000) ‘Net phone creator’s recipe: keep the technology simple’, *Associated Press story*, 16 November.

Kyodo News Service, Japan Economic Newswire, 9 October 1996.

Mainichi Daily News, 21 July 2005, p. 8.

Ministry of Internal Affairs and Communications (2005) *Information and Communication in Japan 2005*. Online. Available: www.soumu.go.jp/johotsusintokei/whitepaper/eng/WP2005/2005-index.html (accessed 25 July 2009).

—— (2010) ‘Chapter 4: Current Stage of ICT’, *Information and Communications in Japan White Paper 2010*. Online. Available: www.soumu.go.jp/johotsusintokei/whitepaper/eng/WP2010/chapter4-1.pdf (accessed 3 January 2011).

Ministry of Public Management, Home Affairs, and Posts and Telecommunications (2004) ‘Information and communications in Japan’, White Paper 2004. Online. No longer available: www.johotsusintokei.soumu.go.jp/whitepaper/eng/WP2004/2004-index.html (accessed 5 October 2005).

Multichannel News International, May 2002, 8(4): 20.

The Nikkei Weekly (2001) 12 November report.

—— (2004) 29 March report.

—— (2005a) 7 February report.

—— (2005b) 16 May report.

—— (2005c) 15 August report.

NTT DoCoMo (2010) ‘DoCoMo takes the next big step with LTE’, *Mobility: DoCoMo Newsletter*, December 24. Online. Available: www.nttdocomo.com/features/mobility31/index.html (accessed 8 February 2011).

nua.com survey (A). Online. Available: www.nua.com/surveys/how_many_online/n_america.html (accessed 5 October 2005).

—— (B). Online. Available: www.nua.com/surveys/how_many_online/europe.html (accessed 5 October 2005).

plusd.itmedia.co.jp News Report. Online. Available: http://plusd.itmedia.co.jp/mobile/articles/0907/01/news074.html (accessed 3 July 2010).

SinoCast China IT Watch (2005a) 29 March report.

—— (2005b) 11 April report.

—— (2005c) 28 June report.

SinoCast Daily Business Beat (2010) 11 August report.

Telecommunications Carriers Association of Japan Telecom Data Book 2010. Online. Available: www.tca.or.jp/english/databook/ (accessed 19 January 2011).

Tom Online Inc. (2005) Presentation, Beijing, July.

White Paper on the Internet in Japan (2002) Online. Available: http://kiko-net.com/Foreigners/White_Paper/White_Paper.php (accessed 25 June 2005).

Xia, J. (2011) 'The third-generation-mobile (3G) policy and deployment in China', *Telecommunications Policy*, 35: 51–63.

Xinhua Economic News Service (2005a) 1 March 2005.

—— (2005b) 24 March 2005.

7 Mobile communication and individualization in Europe and in China

Leopoldina Fortunati and Pui-lam Law

Modernization, industrialization and individualization: a premise

An analysis, even rapid, of the scientific literature produced so far on the mobile phone enlightens the reader on significant themes regarding the diffusion and use of this device, but, among them, we might recall three very key issues: the question of time, the notion of the separation between public and private spaces, and the development of the individualization process. The debate on time has described the mobile phone as the emblem of the zeitgeist (Katz and Aakhus 2002), has discussed the new category of presence–absence (Fortunati 2002; Gergen 2002), and explored the concept of rhythm (Haddon 2004; Licoppe and Smoreda 2005; Ling 2004; Sørensen and Pica 2005), which derives from the temporal qualities of interpersonal relationships (Werner and Haggard 1985). The question of private and public has been debated since the beginning of the discourse on the mobile phone. In fact, this debate has registered the big changes that the use of the mobile phone has provoked in the perception and social definition of what is public and what is private (Raiti 2007). It shows how mobile phone use puts into crisis the division between the public and the private space, which has been one of the key aspects of modernization. Finally, the question of individualization has emerged since the second wave of mobile phone diffusion and adoption, given the rapid transformation of this device from being a mobile technology to a personal technology. This shift in mobile phone identity has accompanied and enhanced the individualization process in Europe. The mobile phone in the everyday life of the individual more than any other machine has meant 'one machine, one individual', as well as the strengthening of the individual communication and information capability. The mass appropriation of the mobile phone is explained by the powerful support that this device has offered to the individualization process, enhancing the sense of power and freedom of the individual.

All these topics have been so far analysed on a micro level, that is, like social micro phenomena connected to the practices of use of the mobile phone. But they are at the same time at the core of fundamental processes, such as industrialization and modernity, and of macro sociological elements, such as the structure of the productive system and social organization. Not by chance, when Marx speaks about machines, does he mean or explicitly speak of the factory system, with its social norms and rules, including the structure of power which makes the traditional social organization homogeneous, uniform and standardized. These three topics (time, private/public and individualization), however, should be analysed also from a diachronic point of view. In fact, to understand the social meaning of these aspects involved in mobile communication, it is worth recalling how they have been produced historically. Only in this way might we be able to outline and explain the differences in these processes in the old industrialized countries in Europe and in the new industrialized countries such as China.

Let us premise that: (1) with the term 'industrialization' we mean here that the economic system implies a differently skilled labour force for the production and distribution of products and that allows an increase of labour productivity and per-capita income; (2) industrialization has always passed through the development and adoption of technology; (3) industrialization is linked also to modernity. By means of machines people are obliged to adapt to a new way to work and to live. The machine, which is based upon a principle of rationality, contributes to create the modern environment that is 'rational', lay and increasingly urban (Veblen 1904).

At a macro-sociological level, industrialization and modernity have marked since their beginnings these three fundamental topics: the importance of time, the separation between the public and private dimensions, and the development of the individualization process. Time, for instance, in the technical environment of urban-industrial societies, has become a very important datum. Mumford (1964: 14) argues that the watch was the key machine of the modern industrial age. This new conception of time has transformed also the notion of duration. If in the rural world the maximum value is given by what lasts in time, in the technical world quality and duration of goods become dissociated (Sombart 1902/1927). Regarding the dichotomy public/private, being modern for the individual means to renounce the protection of an intermediary authority and of powerful people and to accept the separation between the private and the public (Ferrarotti 1970). Even the idea of progress, which is the key idea of the modern world, is developed in an individualistic perspective. Individualization, as we will see later, is a process developed on the basis of the wage system and crosses the various stages of the capitalist system. This process was seen by Marx to be clearly connected to the progressive isolation of individuals (Simmel 1908/1998).

But to better highlight the evolution of these topics (time, private/public and individualization) in the old industrialized countries, for a comparison with what is happening in China we need to outline a little more the main characteristics of industrialization (and modernity). Although we are still lacking a dynamic and global analysis of the industrialization process (Ferrarotti 1970), we know at least three important aspects of the industrialization process (and modernity). The first aspect is that industrialization has represented in Europe a rupture of the socio-economic routine and the source of its legitimacy: the tradition. Making this rupture, industrialization provoked tensions between: (a) the traditional social practices and the new needs of rationalization; (b) the old type of social relationships, in which individuals and groups perceived themselves as friends or enemies, and the new (modern) type of depersonalized and psychologically neutral social relationships required by a society technologically advanced (Sombart 1901/1991); (c) a social structure that changes but that is still unable to be depersonalized, and a rising social structure, which is dynamic and functional, and needs to distinguish between private and public. Industrialization is today well developed also in China. Like in Western countries, the development of industrialization has here broken the previous local socioeconomic routine and de-legitimated its tradition, by creating tensions between the traditional social practices and relationships and the new 'rational' social environment. The problem in China is that one has passed to post-modernity without passing through modernity.

The second aspect is that industrialization seems a kind of self-generating process which shows an internal logic of development (Ferrarotti 1970: 73–77). It is in force of this powerful logic, which is ideologically neutral, that different political ideologies – such as communism – might coexist with industrialization, like in China. Here, capitalism and communism are giving birth to a hybrid social system that should be analysed very carefully. But beyond China, it is increasingly clear how industrialization is becoming a global process. Although countries are divided by different political ideologies and institutional structures, industrialization appears as the characteristic which defines states like Italy or China as modern. In other words, industrialization is a distinctive and fundamental component of the modern world. A country which refuses industrialization refuses to belong to the modern world.

The third aspect is that industrialization represents itself as irreversible. Thanks to new techniques and technologies, the old international division between industrialized and agricultural countries has been overcome. Techniques and technologies have allowed the modification of data which before were unchangeable, such as a lack of materials, and a location far from the traditional centres of production and consumption, resulting in high costs of transportation (Ferrarotti 1970: 74). The structural aspects of industrialization – its being disruptive of tradition, self-generating and irreversible – show the pressure that the perception and conceptualization of time, the social distinction between private and public and the creation of a distance from others are subjected to.

In this chapter, we focus on the individualization process, which has been enhanced, as we already noted, by the diffusion and adoption of the mobile phone in Europe as well as in China, but which is also at the core of industrialization, modernity and even post-modernity. From what we have said so far, it emerges that the individualization process, in addition to increasing the distance of individuals from their community, relatives and family, implies also the depersonalization of the individual, the construction of psychologically neutral social relationships and the clear distinction between private and public. Furthermore, this process is not only at the core also of modernity, but, as we said above, seems irreversible. Our research questions are: Does individualization appear in China to be similar to or different from how the phenomenon has developed in Western countries? And is the mobile phone enhancing this process as it did in the old industrialized European countries? After this short introduction on the tangle constituted by industrialization, technology and modernity, and drawing on the findings that have emerged from many years of studies on the practices of use of the mobile phone in China, first of all we will examine the theoretical and historical framework in which the individualization process has developed in old industrialized countries (Sennett 1994). Second, we will analyse to what extent the characteristics of this process are structural, or are instead cultural and therefore subject to re-semanticization by the Chinese culture. Third, we will discuss the implications of individualization for China, and last of all will proceed to a final discussion of the role of the mobile phone in China towards individualization.

The theoretical and historical framework of the individualization process in old industrialized countries

Let us start our analysis of the individualization process in Europe by describing the framework in which this process has developed. Even a short reading of individualization will allow us to understand better the tissue of historical events and social, economic and psychological phenomena that in time have converged on it. We cannot help recalling one of Marx's most important ideas (1964): the real essence of the capitalist system is above all in the social relations to which it gives life, based on the sale by workers of their labour in exchange for wages. The capitalist social relation thus recognizes the worker's freedom, within certain limits, in exchange for their exploitation. It is a social relation that leads to the progressive weakening of bonds of family members, relatives and neighbours, to the point that in Western history civil progress is linked to the development of an increasing social distance and psychological neutrality of the individual.

The *Magna Carta Libertatum* (15 June 1215) was the premise and at the same time the driving force for the formation, first, of the original accumulation of capital and, successively, of industry. But it was in the eighteenth century that civil society took such enormous strides as individuals began to present themselves freed from those bonds that previously had made them – and these are Marx's words: 'an accessory element to a determined and circumscribed human conglomerate' (Marx 1964: 4). In this context the French Revolution was a further thrust towards the development both of the capitalist system and the process of individualization. To understand this progressive individualization, Marx goes back to Aristotle, who maintained that human beings are not only social animals but are also animals capable of being isolated in society. It is in fact the progressive isolation of the individual, on which their individualization is based, which increasingly incorporates not only freedom but also equality. With the Western capitalist system individuals are made free, in that they are mutually recognized as owners of goods (Marx 1857–58/1970, Vol. I: 213). At the same time these individuals are made equal to one another by the social relations that they are subject to, as entities exchanging equivalents, that is, objects conventionally equal (Marx 1857–58/1970, Vol. I: 209 ff.). According to this framework, the diversity of individuals and their goods constitutes both a reason and a stimulus for their social integration. Obviously this is a democracy designed by the law of exchange and so all within the conservation of the logic of capital. But it is still a 'liberatory' stage in the history of humanity.

So, the development of this mode of production, with its driving centre in the England of the last quarter of the eighteenth century, which appeared following phenomena such as the original accumulation of capital, the end of home manufacture, the chasing of the peasant classes off their lands, and the Enclosures (Marx 1964), was forged by a system of forces, thrusts and counterthrusts that were tightly interconnected. At the same time, the characteristics of its path were determined by the struggles that the European proletariat put up. In the network of freedoms and civil rights created by the social mobilization of wage and non-wage-earners we can discern the subjective behaviour of a variable – the labour force – that over time has fought to be ever more independent (Ferrari Bravo 1975) and to claim an ever greater part of the value produced. Among the various stages of its struggle might be cited the struggle for the reduction of the working day, the conquest of ever higher and more widespread levels of education in the social body, the questioning of authoritarianism (Adorno 1997) and the process of emancipation/liberation of women and children (Dalla Costa and James 1973). Within the process of individualization, many phenomena can also be discerned. To be very schematic, among the more temporally remote objective phenomena, we may mention the formulation of democracy in various historical contexts and with various structures, of which the most well known is ancient Greece, the conception of the Roman *civis*, and the experience of Italian 'communes'. Also the diffusion over a wide scale of Christianity (every human being has a soul) helped to develop the concept of the individual, by stressing personal responsibility in the face of action. Another important social phenomenon connected with the development of industry is urbanization, due to the abandonment of the countryside and to internal and external migrations (Marx 1964). Also this phenomenon enhanced the progressive thinning and weakening of links and micro-controls that these implied within the family. While industrialization – as we said – progressively enabled individuals to reach economic independence, urbanization ensured also the lonely individual at least a certain sense of company, which was procured by the anonymous, but lively, crowd of the urban landscape (Simmel 1908/1998).

The interpersonal relations that were the basis of day-to-day living, founded on close knowledge and a certain kind of social ritualization, but also on strong mutual control, have gradually weakened. In the big towns people felt the heaviness of solitude, anonymity and spatial extremity (Riesman *et al.* 1950), but also freedom. At the same time, inside the new families that have been created, the process of weakening family ties has quickened, as the weight of the original family and the power hierarchy are also weakened. The social division of work and competences between men and women has given way to less differentiation of gender. Women in particular have earned access not only to levels of education equal to if not higher than those of men but also have had in many cases access to extra-domestic work and so to their own income. Women and young people, children included, have been the social subjects with the most dramatic changes in behaviour at a mass level. Emotions, sexuality and love relations have progressively come into the limelight, as the aura of secrecy, tacit or implicit, that surrounded them and which made their social negotiation quite difficult, has begun to dissolve under the blows above all of women and young people. They have therefore started to manifest themselves in more expressive forms, more declared and therefore negotiated. The regime of 'double morality' as also the configuration of the various social roles has become with time more and more flexible (Sennett 1994) and multiform, giving rise to a plethora of figures. For example, as regards the man–woman relationship, apart from being fiancée and wife, the woman has also become the girlfriend, friend, companion, mistress, or just cohabiter.

The direction taken by the masses in Western countries has been, without doubt, towards autonomy, responsibility, the personal freedom of all members of society, including women and the elderly, as well as on the part of the individual from other members of the community and society (Sennett 2003). The cost paid has been that of strong fragmentation, alienation and desocialization, and not least, a loss of political power on the part of the various classes. In other words, that process of the progressive isolation of the individual has proceeded inexorably, but which, however, corresponds to a personal and social identity that is better articulated than in the past. In this context, communication has taken on an ever-greater importance, because it has become the means by which social workers, women and young people have negotiated their individual rights and mediated their respective interests within them. But more generally when the masses no longer saw force as a means of changing their conditions of life, the word and voice took on a fundamental role. It is with these new tools that the masses built up their political importance and their new social roles. So, the word and the voice not only became crucial places of negotiation between single individuals or classes, they also became essential places of public opinion and political bargaining. The information and communication technologies found themselves mediating the complexity of communication itself, which is made up of word and silence, of play and evasion, of multiple meanings and so on. And it is in this tissue of possibility, potentiality and negotiation that the technologies reveal their real social meaning in regard to individualization. This meaning depends on the logic that inspires their use. In certain moments the economic system makes them work in order to build the standardized individual; in others people manage to use them to construct their independence from the plans of capital.

To conclude this line of reasoning, in Western countries the individualization process has been a very complex process made up of several events, political and economic processes, new values and attitudes, cultural dimensions, collective behaviour, and social phenomena. The modern individual was founded on personal freedom, economic and political independence, responsibility, equality, a network of civil rights and a much stronger personal and social identity. The modern individual became more urban, more educated, more emancipated, mainly Christian, a communicative and emotional subject, a democratic citizen. The dark side of this process was fragmentation, isolation, alienation and desocialization. But at the basis of the individualization process there is, however, social protagonism as an independent variable of capitalist development.

With the evolution of the capitalist system in the West, all the powerful economic, social and political changes which have been triggered resulted in a negotiation that took place in the social context from which this same economic system in a certain sense started. In other words, the 'capitalist system' was born out of Western culture, from its social structures and from the Christian religion in its variants (cf. Weber 1965). Now, it is important to understand which configurations and characteristics of this process are structural, intrinsic to the process itself, and which are instead social and therefore may be re-semanticized by Chinese culture.

Migration of capital, industrialization, and the individualization process in China

The mobilizations of the various classes in the West in the last 40 years have developed such a political power as to force capital to migrate from the old industrialized countries to China (as well as to other developing countries), conditioning its penetration there, speeding it up, slowing it down, deviating it, interrupting it (Gambino 1975: 318). Globalization must be seen not only as a response on the part of capital to this very powerful cycle of struggle that has overwhelmed the West, but also as an attempt to re-affirm capitalist supremacy at a world level. In other words, globalization represents not only the victories but also the limits of the struggles of social subjects in the old industrialized countries. Here, there was no way left to raise the margins of profit, as the cost of labour, following worker and social struggles, was too high in respect to the claims of the entrepreneurial class and had been blocked for too long by trade union and legislative defence barriers. The solution was to move capital to those countries where the less developed conditions of life offered more advantages than disadvantages, the first of which being that of a much cheaper labour supply which was outside generally accepted trade union and institutional protection.

As yet, there have not been any important attempts on the part of workers' movements and organizations either to block the de-industrialization of Western countries or to prevent the flow of capital and the development of the capitalist system in countries that only yesterday were mainly rural. This massive migration of capital towards China has taken place practically unhindered, given the lack of internationalization of the political associations of the workers' class and in general of the various political subjects. It was this very lack of proper cosmopolitanism on the part of social movements and class organization that allowed this kind of capitalist response: the migration of capital towards those areas and countries where the cost of labour was low in comparison.

The development of the empire, as Hardt and Negri (2000) call it, has ended up creating on a world-wide scale a higher level of economic integration in production, distribution and finance, and a wider level of integration of industrialized countries, those in transition, and those newly developed, like China. An important aspect stressed by many observers is the speed of monetary transactions, goods and individuals present today in China, where the concentration of capital on the southern coast of the country has managed to shift almost 140 million Chinese peasants, men and women, who have abandoned the villages where they were born to go in search of an income in the factories in the south.

The first result of this migration of capital in China is the creation of the world's largest workforce which by 2008 numbered more than 225 million. China is considered today the factory of the world. The capitalist system and the wage relations exported from the West in a few years are radically changing the structure of Chinese society. The modern Chinese individual is shaped by economic independence and urbanization. The millions of male and female workers, migrants or otherwise, who earn a wage obviously make up a vast laboratory of 'individualization of ownership': wages are a social relation that involves millions of Chinese men and women, who have direct access to resources for survival. So, the economic basis for the development of the individualization seems to be already spread among a consistent part of the Chinese population. But the political, social and psychological determinations such as freedom, personal responsibility, equality, civil rights and democracy that in Europe have accompanied this economic basis have a quite different meaning in contemporary China.

Regarding the traditional social structures which surround the individual, like the family, matrimony, sexuality, etc., the use of the mobile phone seems to follow the current impetuous capitalist process in China that is producing a hybrid society (Yuen *et al.* 2004). In fact, even if what the capitalist system generally produces is modernization of the production process, in China the development of the process of capitalist production, at least for now, does not necessarily imply any great modernization of social structures, relations and institutions. Actually it is better for the social structures of the past that still survive not to be questioned, as they can soften the too rapid changes that are taking place inside the industrialization process. These structures are, on one hand, the fruit of a long historical tradition and, on the other, of an unfinished revolution, in that for example the laws promulgated by the Communist Party to obliterate the inequality of relations between men and women have been so ineffective. However, we cannot help but note that these ancient social structures are being questioned by new generations with more strength. In fact, among young couples a greater equality between husband and wife is now more common, in the same way as there is quite obviously a kind of individual autonomy emerging.

According to a research project carried out in 1995 by the Institute of Scientific Research for Family Planning in Shanghai on a sample of 2,580 unmarried young people, 63.9 per cent of the girls stated they had had sexual relations at least once, 39 per cent were pregnant and 89.5 per cent of these had decided to abort (Aresu 2000). These figures give an idea of how sexual behaviour is changing today among young Chinese people. Other research carried out by Law and Peng (2006) shows that the use of the mobile phone discourages in some cases the maintenance of tight relationships with the family of origin.

But other historical-political factors might have a special influence on the individualization process. The first factor is the different political structure of power. Despite the enormous capacity for adaptation to capitalist penetration demonstrated by Chinese society, the penetration of capital into China has found a society that had been living for the last few centuries in a very special kind of economic and social situation. Adriano Sofri (1974) refers to the Marxist concept 'Asian mode of production' to describe a mode of production that had been characterized by strong continuity, stability and immobility. This mode of production was based on a rather flat political-organizational form of society, in which absolute power on the part of the Emperor was justified also by the necessity to promote public works, both hydraulic (dams, irrigation channels and navigation canals, water mains) and infrastructural (such as the construction and maintenance of roads and fortifications like the Great Wall of China), as well as monumental architectural works (palaces, tombs, etc.). In Europe, the political and social pyramidal hierarchy was born from the Greek tradition, in which politics was conceived (by Aristotle, for example) as the commitment to the pursuance of one's interests when these coincided also with those of others. From here was born the tradition of a government that is the mediation of different interests and the creation of consensus. This contrasts with the long-lasting absolute power that is found in China.

In the West the process of conquest of the individual's freedom came about by means of the experiences already described: Greek democracy, the Roman *civis*, habeas corpus, Italian communes and the French Revolution. These experiences have created over time an ever-broader base of access to rights of citizenship, the right to consensus, but also dissent and criticism, with which those in power in the West have always had to come to terms. In the West, it was the wage structure that has been the main base that led to the construction of autonomy and personal individualization. Starting from the social relations that were the basis of the wage structure, the division and organization of work and power in the social pyramid probably constructed the sense of an infrastructure that threw its shadow over the social imagination and also inspired the technical infrastructure. This existed more widely in the West than in the East because an organizational connective tissue had been created in the West which is the basic structure of any system that should work in a modern way. In China several observers talk about the current formation of a political-technocratic power, which maintains the characteristics of the old absolute power. The individualization process in China, although proceeding with the diffusion of the wage relation, might encounter more difficulty in the related process of empowering individuals. Whereas the modern individual, who was founded in the old industrialized countries characterized by personal freedom, political independence, responsibility, equality and a network of civil rights with a much stronger personal and social identity, became a more educated, emancipated, communicative and emotional subject with democratic sense, this is just not the historical past and tradition in China.

In China, in the periods of Spring and Autumn in the warring states (600–300 BC), a hundred schools of thought developed, and scholars were at liberty to travel around and test their theories in different states. The Qin state conquered all the other states and ended the warring states period in around 220 BC, but 80 years later, the Qin was replaced by the Han dynasty. At some time around 140 BC, Confucianism was espoused as the chief school of thought. Taoism, legalism and others were considered secondary and unimportant; Buddhism had not yet entered China. The period of pluralistic thought and ideas had come to an end and, since then, Confucianism has dominated Chinese society. The Confucian five classics were the required texts for the imperial examination, through which officials were selected. The five classics then evolved into thirteen classics and a great number of classic annotations in the Sung dynasty (AD 960–1279). Inevitably, Confucian ideas influenced state policy. Confucianism also influenced the grassroots; those who failed the imperial examination would then become teachers in their province, at the county or the village levels. Confucianism became pervasive in all levels of Chinese society.

Confucianism, as a set of ethical ideals, had provided justification for *li*, the traditional rites, among which family values were considered central and hence regarded as the primary basis of the social norms. Of the family values, filial piety, meaning respect and obedience, is the primary value, and from this concept the five cardinal social relationships were developed: ruler and subject, father and son, husband and wife, elder and younger brothers, and friends. Simply put, these relationships have defined the basic social fabric of Chinese society and they have also set the underlying structure of patriarchal familism in China. This occasioned a patriarchal hierarchical power relationship and male domination in Chinese society (Yuen *et al.* 2004).

This basic social structure prevailed until the Chinese Communist Party came to power in 1949. The Chinese Communist Party attempted to transform traditional social values and the basic social fabric, and several revolutionary movements were put in place. The formation of agricultural co-operatives and then the commune system were designed to abolish the feudal system. On the surface, these new organizational entities at a social level appeared successful as the patriarchal system was suppressed and rules stipulated that women should be free from the control of men. But after the implementation of the economic reform and opening-up policy in 1979, traditional practices saw resurgence. The deep structure has remained intact even though the state promulgated policy to uproot it (Potter and Potter 1990). At the early stage of the economic reform, the suppressed patriarchal influence on daily matters and the domination of women made a comeback.

The second factor is that, whereas the encroachment of globalization and the development of the capitalist system in China are changing this country profoundly, this process is also changed profoundly by China in turn. The modelling of the social system in the West has become an integral part of the new processes of accumulation in China, whatever the model of labour relation that Chinese social formation has previously imagined. The development of industrialization in China is a laboratory of social experimentation that begins to change the deep social structure significantly. If the capitalist attempt was, by migrating to China, to bring the cost of labour down to a much lower overall price, this attempt is already subject nowadays to mass negotiation for higher wages (Law *et al.* 2005). Today, migrant workers have shed their loyalty to the firms for which they work. If they discover that a firm is paying higher wages, even it is 300 kilometres away from the spot, migrant workers are ready to move there immediately. After all, workers who have been travelling 2,000–3,000 kilometres to find a job obviously do not hesitate to go another 300 for a better wage. On this issue – a better wage for workers – there is a lively debate now in China at a political level, which also takes place in the mass media.

The third factor is connected to the great internal migrations. China is the only Asian country where there has been so massive and rapid a development of both industrialization and urbanization, thanks to the arrival in the areas of urban drift of masses of migrant workers. China is a special case in comparison to other developing countries, where peasants are abandoning their villages to migrate to a metropolis where often they are obliged to make use of only the infrastructure, the essence itself of its urban structure. In China, instead, millions of people coming from the most out-of-the-way places in the provinces are pouring into the cities and industrialized areas of the south to become industrial workers. This unprecedented scale of migration is dramatic testimony to the crumbling of family ties and the constant growth in numbers of workers, men and women, who live alone, at great distances from their parents and families. They suffer solitude in the collective dimension of the dormitories where thousands of migrants find shelter. But at the same time this means that they are living far away from the micro-control which their families exercised on them. Urbanization, however, involves not only workers, but also students. University studies are in fact another reason for young people moving to towns and being freer as they stay far from the family for their whole period of study. Uprooting students from the family and therefore from the control of family and relatives is also helping to transform relations between men and women among the young generations. But, more than that, in general, millions of Chinese families are experiencing an internal separation among their members: parents divided from children, husbands from wives. This has brought about a significant impact on the traditional social structure. The patriarchal influence has been diminishing because of the distance. In addition, workers' (referring to both male and female) economic independence has also had radical consequences on the power structure of the family (Cheng 2011; Thireau 1988).

Individualization and the mobile phone in China

In this new Chinese capitalist context that sees a hybrid development of the individualization process, what role does the mobile phone play? This device should in general play, like all the other communication and information technologies, a powerful role of creation of value.

> Capital tends to overcome every spatial obstacle. The creation of the physical conditions of exchange – that is, means of transport and communication – therefore becomes a necessity for it, but in a completely different measure – it becomes the annihilation of space by means of time.
>
> (Marx 1857–58/1970, Vol. II: 161)

Sombart (1901/1991: 52–53) might help us to better understand the role of the mobile phone in respect to the individualization process. To him, the evolutionary principle of technology is freedom, in that it emancipates human beings from the limits of the organizational process of the nature to which they are constricted. But this technological principle ends up being in sharp contrast with that of economy, which is the lack of freedom. Now, the use of technology in capitalist societies grafts together paths of individual and social freedom, which the economic system attempts to shape and make rigid so as to maintain the 'modern' slavery of wage labour. From Sombart's analysis emerges a further invitation to see technologies as acting within the logic of capital and therefore to see them really as spheres of valorization (returning again to the definition given by Marx). So, on the basis of the discourse of Marx and Sombart, the mobile phone could be seen as a potentially liberatory technology from the constriction of labour and social organization, but in any case always functioning as a tool for valorization.

In the old industrialized countries as well as in China, the mobile phone enters this working process as a device which is used to support all the activities that can be included under the umbrella of reproduction labour (Fortunati *et al.* 2008c). This labour is made up partly of material labour, but also increasingly of immaterial labour: organization, communication, information, entertainment, civil and political engagement, education, interaction, affects, love, sex. The material and immaterial reproductive labour converge to create a plus-value that is embodied in human beings. This notion of domestic or reproductive capital includes, systematizes and enlarges the notions of human, social and cultural capital. In this framework, the mobile phone, like the other information and communication technologies, works on the side of immaterial reproduction. In other words, it helps to rationalize the immaterial dimensions cited above.

The double aspect of the mobile phone as a potentially liberatory technology and as a tool for valorization has structural and social consequences in the process of individualization in China. The social empowerment that Chinese obtain through the mobile phone is undeniable and from this point of view the mobile phone acts in China as a 'technology of freedom' (De Sola Pool 1988), possibly benefiting also the enlargement of democracy (He 2008). At the same time, the mobile phone is acting as a valorization tool which accompanies the new social stratification which is reshaping Chinese society. The cost of its use arises against the same weak social subjects, as these people spend a large part of their salary to pay for their device and its use. Of course, they are obliged to limit their personal mobile calls for economic reasons. Despite this self-limitation, their mobile expenses remain very high and bind them within the lowest strata of the Chinese population. This is the contradictory situation that the metaphor of the mobile phone as an empty shell of contemporary modernization for these social strata describes (Law and Peng 2007).

But the practices of mobile phone use which contribute to develop individualization are enlightened also by several other social processes or phenomena. For reasons of space, we decided to analyse only some of them.

1 The recent development of urbanization in China has made 140 million Chinese citizens more isolated than in the past because they are far from their family, relatives and their community of origin. From a recent research project (Fortunati *et al.* 2008a, 2008b, 2009), it emerges that especially the weakest portion of the 487 respondents in Beijing (people born in the countryside, with low education and income) perceive that, thanks to the mobile phone, their connectivity and communicative and social networks not only at a quantitative level, but also at a qualitative one, have improved.

2 China has a huge territory, which might create in its citizens a strong sense of anomy. An important part of Chinese territorial identity has always been connected also to the region where the individual was born. So the mobile phone has been the ideal device to reinforce the self by supporting the maintenance of links and social relationships with people from one's own town and region of origin, in addition to developing relationships with people in the new town and further develop new social relationships in mobile cyberspace. This device was also the ideal tool to achieve the annihilation of space by means of time (to use Marx's words), which is so useful in China, given the size of its territory.

3 The fact that the main information on the labour market (new jobs, wage levels and the like) is spread via the mobile phone makes access to this device a must for people who wish to change jobs or to improve their salary. The mobile phone user becomes the centre of a network that is able to give and receive precious information. The free circulation of information through the mobile phone is an important condition for the worker to get greater autonomy. When the mobile was not ubiquitous, workers had to obey whatever demands were made by the capital system. It was not infrequent that female workers were forced to sell their bodies in order to survive in the factory. Some had to be submissive to the factory management when faced with invidious treatment. Nowadays, however, mobile phones, to a greater extent, have empowered migrant workers to tackle unjust circumstances and liberate them from oppressive situations that they have faced in the host cities (Qiu 2009). Undoubtedly, this has contributed to individualization.

4 Chinese people probably use the mobile phone, and especially the Short Message System (SMS), in order to overcome the rigidity and the emotional censure connected with the communicative procedures in body-to-body encounters (Fortunati *et al.* 2008a). The spread of this device is explainable only if we keep in mind that, thanks to this facilitator of interpersonal communication, Chinese have been able not only to survive in the industrialization and modernization processes but also to increase their *guanxi* (interpersonal connections). In the past the self-sufficient Asian communities suffered especially from a lack of communicative infrastructure. This lack in its turn helped to consolidate their isolation, constituting an essential part of an unaltered continuity (Marx 1857–58/1970, Vol. II: 162). In the last decade users have assigned to the mobile, more than any other communication and information

technology, the role of communicative 'infrastructure' *par excellence* and used it to support individualization by mitigating the progressive social isolation of each one, starting above all from immigrant workers and continuing in general with people who recently have become urban.

5 There is resistance to modernization, opposed by the deep social structures, as we already mentioned. In regard to this issue mobile phone use in some cases supports these social structures, but in other cases is disruptive and for this reason is supporting the individualization process (B. Yang in this volume, Chapter 9). As shown in a study (Yuen *et al.* 2004), the familial culture continues to be based on the five fundamental bonds: sovereign and subject, father and son, husband and wife, among brothers, and among friends. These relationships go on to affect all other social bonds, partly because the lack of health assistance, adequate social services (kindergartens, etc.) and an efficient pensions system make intra-generational family solidarity necessary. Of these bonds, the father–son dyad, according to Francis Hsu (1968), continues to be the crucial one, because it affects all the other kinds of bonds. It is evident that this kind of family still lies at the core of Chinese society, even if ritualization of family practices is weakening. In this context, the three authors maintain, the condition of women is even less advanced than in the recent past in China, because man–woman relations continue to be based on the notion of male superiority and female inferiority. This notion is so strong that it even crosses intergenerational bonds, which are based on the authority of elders. Indeed, the status of a man is higher than that of an elderly woman. At the same time, the organization of social relationships is based on the father–son dyad, which implies 'continuity, inclusiveness, authority and asexuality', and shows on one hand that the structure is configured as stable and not subject to change due to internal or external factors, but on the other that there is also an infantilization of youth and adults. If anything, with industrialization, the oppressive model of bond between men and women 'has shown signs of becoming more entrenched' (Yuen *et al.* 2004: 9). We are assisting in an adaptation of the capitalist system to authoritarian social structures, which are functional, as long as they last, to the social discipline required by this phase of development in China. In this framework, sometimes Chinese use technology in general, and the mobile phone in particular, in order to subvert this adaptation, but at other times use the mobile phone to improve it.

The research so far conducted on the social role of the mobile phone in China (Law and Peng 2006; B. Yang in this volume, Chapter 9) seems to show that people use the mobile phone not only to keep in contact with a family broken up by migration, but also to build up an individual sentimental and affective freedom in love relationships. Therefore individualization is supported by the diffusion and use of the mobile phone, as well as by the growing consumption of goods of exchange, with the inevitable consequence of the proliferation of tastes and personal and group inclinations.

Conclusion: towards a Chinese modernity?

Many scholars have observed, as we already reported, that modernization is today less and less accompanied by Westernization (Cacciari 1994). This is true, although it is less true in some cases such as fashion, which is therefore one key element in the presentation of the self and so in the making of individualization. People probably choose to wear Western clothes because this style gives them the sense of modernity and of a modern self. In the same way Simmel has suggested that we see fashion as a vehicle of modernity, we might suggest that it should also be considered the emblem of post-modernity and globalization, seeing that, together with information, it manages the present, by giving a rhythm to social life and also marking the passage of time by continually introducing changes.

But, apart from this, from the data available, it seems evident that the diffusion and use of the mobile phone have been made to work within the processes of individualization and social re-engineering. China is symptomatic of how the new world order is shaping economically and of how the principle of fusion is the basic principle underlying the workings of globalization. The fusion of modernization and post-modernity is very obvious in China. As we have seen, there is a shift to industrialization, aided by internal migration, but there are also signs of a shift to the development of the service and information sectors (think of the almost 100 million computers present in China today). Actually here more than anywhere else we are seeing a hybridization of the modern and post-modern.

Individualization in China seems not to be a replica of what has happened in Western countries. There is an extensive negotiation among citizens, social structures and institutions on the one hand, and the logic of capital on the other, which is leading to an important realization of the cultural and social significance of the individualization process in China (Geertz 1973). The development of the latter in fact is more difficult in China, because traditional Chinese thought has little room for the concepts of individuality or privacy (Law and Peng 2006; Whitman 1985). The so-called 'technologies of freedom' (De Sola Pool 1988), which have found such a favourable environment for their incubation and diffusion in the Western world and where they have served to unlock individualization (especially the mobile phone), are domesticated, as we saw, in 'other ways' in the Land of the Dragon.

References

Adorno, T. (1997) *La personalità autoritaria*, Rome: Edizioni di Comunità.

Aresu, A. (2000) *L'educazione sessuale in Cina: la mostra 'nella primavera della vita'*. Online. Available: www.tuttocina.it/Mondo_cinese/111/111_ares.htm (accessed 2 January 2008).

Cacciari, M. (1994) *Geofilosofia dell'Europa*, Milan: Adelphi.

Cheng, C.T. (2011) 'Floating workers and mobile QQ: the struggle in the search for roots', in L. Fortunati, R. Pertierra and J. Vincent (eds) *Migrations, Diaspora, and Information Technology in Global Societies*, New York: Routledge.

Dalla Costa, M. and James, S. (1973) *The Power of Women and the Subversion of the Community*, Bristol: Falling Press.

De Sola Pool, I. (1988) *Technologies of Freedom*, Cambridge, MA: Cambridge.
Ferrari Bravo, L. (ed.) (1975) *Imperialismo e classe operaia multinazionale*, Milan: Feltrinelli.
Ferrarotti, F. (1970) *Macchina e uomo nella società industriale*, Rome: ERI.
Fortunati, L. (2002) 'The mobile phone: towards new categories and social relations', *Information, Communication & Society*, 5(4): 513–28.
Fortunati, L., Manganelli, A., Law, P.L. and Yang, S. (2008a) 'Beijing calling … modernisation and the social effects of new media in China', *Knowledge, Technology, and Policy*, 21(1): 19–27.
—— (2008b) 'The mobile face of contemporary China', *Telektronikk*, 2: 57–67.
—— (2009) 'Practices of mobile phone use in Beijing', in L. Stillman, G. Johanson and R. French (eds) '*Communities in Action: papers in community informatics*, Newcastle upon Tyne: Cambridge Scholars Publishing.
Fortunati, L., Lee, F. and Lin, A. (2008c) 'Introduction to the special issue on "mobile societies in Asia-Pacific"', *The Information Society*, 24(3): 135–39.
Gambino, F. (1975) 'Composizione di classe e investimenti diretti statunitensi all'estero', in L. Ferrari Bravo (ed.) *Imperialismo e classe operaia multinazionale*, Milan: Feltrinelli.
Geertz, C. (1973) *The Interpretation of Cultures*, New York: Basic Books.
Gergen, K. (2002) 'The challenge of absent presence', in J. Katz and M. Aakhus (eds) *Perpetual Contact: mobile communication, private talk, public performance*, Cambridge: Cambridge University Press.
Haddon, L. (2004) *Information and Communication Technologies in Everyday Life: a concise introduction and research guide*, Oxford: Berg.
Hardt, M. and Negri, A. (2000) *Empire*, Cambridge, MA: Harvard University Press.
He, Z. (2008) 'SMS in China: a major carrier of the nonofficial discourse universe', *The Information Society*, 24(3): 182–90.
Hsu, F. (1968) 'Chinese kinship and Chinese behaviour', in P.T. Ho and Tsou Tang (eds) *China in Crisis, vol. 1 bk 2: Chinese heritage and the commercial political system*, Chicago: University of Chicago Press.
Katz, J. and Aakhus, M. (eds) (2002) *Perpetual Contact: mobile communication, private talk, public performance*, Cambridge: Cambridge University Press.
Law, P.L. and Peng, Y. (2006) 'The use of mobile phones among migrant workers in southern China', in P. Law, L. Fortunati and S. Yang (eds) *New Technologies in Global Societies*, Hackensack, NJ: World Scientific.
—— (2007) 'Cellphones and the social lives of migrant workers in southern China', in R. Pertierra (ed.) *The Social Construction and Usage of Communication Technologies: Asian and European Experiences*, Manila: The University of the Philippines Press.
Law, P.L., Peng, Y. and Fortunati, L. (2005) 'Mobile communication, mobile relationships, and migrant workers in Guangdong', paper presented at the international conference France-Chine. Migrations des pensées et des technologies, Guangzhou, China, June.
Licoppe, C. and Smoreda, Z. (2005) 'Rhythms and ties: towards a pragmatics of technologically-mediated sociability', in R. Kraut, M. Brynin and S. Kiesler (eds) *Domesticating Information Technologies*, Oxford: Oxford University Press.
Ling, R. (2004) *The Mobile Connection: the cell phone's impact on society*, San Francisco: Morgan Kaufmann.
Marx, K. (1857–58) *Grundrisse der Kritik der politischen Ökonomie* (Robentwurf); trans. It. (1970) *Lineamenti fondamentali della critica dell'economia politica*, 2 vols, Florence: La Nuova Italia.
—— (1964) *Il Capitale*, Libro primo, Rome: Editori Riuniti.

Mumford, L. (1964) 'Authoritarian and democratic techniques', *Technology and Culture*, 5(1): 1–8.
Potter, S. and Potter, J. (1990) *China's Peasants: the anthropology of a revolution*, New York: Cambridge University.
Qiu, L. (2009) *Working-class Network Society: communication technology and the information have-less in urban China*, Cambridge, MA: MIT Press.
Raiti, G. (2007) 'Mobile intimacy: theories on the economics of emotion with examples from Asia', *M/C Media and Culture*, 10(1). Online. Available: http://journal.media-culture.org.au/0703/02-raiti.php (accessed 10 December 2010).
Riesman, D., Denney, R. and Glazier, N. (1950) *The Lonely Crowd: a study of the changing American character*, New Haven, CT: Yale University Press; (1999) *La folla solitaria*, Bologna: Il Mulino.
Sennett, R. (1994) *Flesh and Stone: the body and the city in Western civilization*, London and Boston, MA: Faber and Faber.
—— (2003) *Respect in a World of Inequality*, New York: Norton & Company.
Simmel, G. (1908) *Soziologie. Untersuchungen über die Formen der Vergesellschaftung*, Leipzig: Duncker & Humblot; trans. It. (1998) *Sociologia*, Turin: Edizioni di Comunità.
Sofri, A. (1974) *Il modo di produzione asiatico. Storia di una controversia marxista*, Turin: Einaudi.
Sombart, W. (1901) *Technik und Wirtschaft*, Dresden: Zahn und Jaensch; trans. It. *Tecnica ed economia*, in T. Maldonado (ed.) (1991) *Tecnica e cultura*, Milan: Feltrinelli.
—— (1902/1927) *Der moderne Kapitalismus*, Munich and Leipzig: Duncker & Humblot.
Sørensen, C. and Pica, D. (2005) 'Tales from the police: rhythms of interaction with mobile technologies', *Information and Organization*, 15: 125–49.
Thireau, I. (1988) 'Recent changes in a Guangdong village', *The Australian Journal of Chinese Affairs*, 19/20: 289–310.
Veblen, T. (1904) *The Theory of Business Enterprise*, New York: Charles Scribner's Sons.
Weber, M. (1965) *L'etica protestante e lo spirito del capitalismo*, Florence: Sansoni.
Werner, C. and Haggard, L. (1985) 'Temporal qualities of interpersonal relationships', in M. Knapp and J. Daly (eds) *Handbook of Interpersonal Communication*, London: Sage.
Whitman, C. (1985) 'Privacy in Confucian and Taoist thought', in D. Munro (ed.) *Individualism and Holism: studies in Confucian and Taoist values*, Ann Arbor: Center for Chinese Studies, The University of Michigan.
Yuen, S.P., Law, P.L. and Ho, Y.Y. (2004) *Marriage, Gender, and Sex in a Contemporary Chinese Village*, New York: M.E. Sharpe.

Part III

Techno-social interplay

Cultural and identity tension

8 Mobile networks and civil society

Philosophical reflections on organizing social participation in the context of Chinese society[1]

Silvia Elaluf-Calderwood

Introduction

The present chapter aims to discuss how mobile networks are being used by individuals (as social citizens) for joining strategically political schemes, aimed at enhancing their social participation in civil society. This discussion reflects on a theoretical framework drawn from the philosophical arguments of the conceptual *Dasein* of Heidegger's *Being and Time* (Heidegger 1962) as well as from other postmodern discourses presented by Castells (2001, 2004), Giddens (1984), Habermas (1984a, 1984b) and Myerson (2001).

We find in the philosophical ideas of *being in the world* and *being with others* an aid in understanding the role of social networks in the life of an individual and the society that person lives in. For the postmodern citizen some of those concepts can also be applied to political participation (Inwood 1997). There is a basic human need to establish social links, either as a participating citizen or as a leader of the masses. Myerson has associated the philosophical concepts of Heidegger and Habermas to the postmodern societies trapped in paradoxical technological situations as in the case of Chinese society, in which the very old coexists with the very modern or postmodern. This coexistence leads to social contradictions or extremes on the social dynamics.

This process of coexistence between the old and the very new technologies is sourced from the generalization of the use of mobile phones across different social spheres (Castells 2001; Rheingold 2002); an evolution of how individuals define their role as postmodern citizens is being shaped in the social world (Bauman 2001). This ontology for example can be applied to the use of mobile phones as a source of temporal contact between people, developing a physical expression of this idea of being able to be in more than one place simultaneously (Myerson 2001). The mobile phone's influence on the perception of being in the world is at different stages of progress in different geographical areas. For example, Asia is ahead of Europe in these changes, helped by the interactivity of the mobile devices available and the extensive social networks that rely on those devices for communication (Berry *et al.* 2003). For a long time, Asian people have extended their ordinary perception of the boundaries of their everyday life due to the association of this life with the use of the mobile phone. There is a perception of perpetual contact as an everyday way of life; this seamless contact is changing the way individuals reach other people with common or opposite interests, hobbies, political activities, etc. (Katz and Aakhus 2002). The idea of reachability and accessibility through the use of mobile phones in this context can be described as giving a wider meaning to the expression 'being in touch anywhere, anytime' (Kleinrock 1998) to an approach that is contextualized and temporal, such as communication happens 'somewhere, sometime'.

Mobile networks reflect upon social networks, since those mobile networks can be understood as important social media when extending the social networks associated with individuals (colleagues, friends, family, etc.), which are used to express their individual or collective views with regard to a political situation (*the moods*). However, the role of the individual in a society is a constantly changing process, and the mobile phone, as well as other media – such as the Internet – is used to exchange information either through discussion forums or actively organizing participation in many types of social interaction, one of them being political activity (Castells 2001).

The sustainability of such philosophical views with regards to the individual, the social networks and the use of the mobile phone in such a political context can only be possible by linking those concepts to social interactions within the frame of the *Dasein*. The new media (Internet, mobile networks, social networking tools) when used as source for informational politics open a window of reflection on how this *Dasein* concept provides a framework for understanding the role these media have on changes related to political engagement. Embedded in the informational politics activities are the individual strategies used by each of us to define the roles and actions to be taken on a particular political context. However, there is a cultural role on this choice of strategies, and this chapter aims to benchmark the Western concept of *Dasein* applied to the socially constructed reality of political China.

We take as a case study a demonstration in 2003 against the Iraq War. The empirical data in this chapter is based on a case study observed during the organization of a public demonstration in front of the US Embassy against the Iraq War by members of the foreign community in Beijing (China) in April 2003. For this demonstration the organizers relied heavily on established social networks (friends, family, associates, etc.) to establish the group organizer and commander of the political event. This group of expatriates are a good example of a social network, which is extended inside China and to the outside world, with highly social dynamics and wide access to multimedia communications. This social network, however, coexists with the Chinese society, with traditions of their own mixing with these new social trends. We discuss briefly how the postmodern citizen relates to the experience of *being in the world* (the Iraq War), *the moods* (the desire to express displeasure/disagreement with the war) and *being with others* (the demonstration itself) and the use of mobile technology and other media.

Finally this chapter presents some reflections on how the use of mobile technology mediated in organizing this specific type of political event. I want to argue that the phenomenology of the observed leads to an explanation for the driving force of the individual desire of *being with others* in the world. These ideas can be extended to the way individuals see their civil role enhanced in virtual political spaces that were not available before mobile networks, spaces in which individuals are able to express their social or political views.

The individual, the social networks and the mobile phone

For the sake of clarity the author wishes to define the term postmodern as used in this chapter. The idea of postmodern (individual and social) must be understood in association with the postmodern world defined by those aspects of contemporary art, culture, economics and social conditions resulting from the unique characteristics of the late twentieth and early twenty-first century. These characteristics include globalization, consumerism, splitting the roles of civil authority, mass communication, and the commoditization of knowledge. Further on, in the world of mobile phones, postmodernism is expressed by the idea of 'talk' as per mobile talk or talk-time, standing for an old sense of person-to-person presence and contact (Myerson 2001). This expressiveness given by talking does not exclude alternative modes of communication.

Heidegger, in 1927, said: 'Man shows himself as the entity which talks'; in those words is reflected a philosophical message for humanity in the act of talking. For Heidegger, 'talk' or 'talking' is a primary activity by which people express their experience of the world (*being in the world*), the experience to be with (*being with others*) and the situatedness of being (*the moods*).

For the postmodern individual the world of the *Dasein* is not limited to the philosophical spheres of thought. It is in a more generic term for the world that surrounds the individual; a world in which the understanding and interpretation of the phenomenological aspects of life take a route or routes in which language and consciousness are used to express temporality, transcendence and freedom. Mobile phone usage provides real and virtual dimensions to the notion of being in each mobile user.

These philosophical concepts are contained in every aspect of the use of mobile phones and mobile networks. Mobile phones make us eager to talk more, and when doing so, the world that we live in is closer to us and we feel that we are part of it. Our individual experiences of the world are exchanged through dialogue about ideas, opinions and arguments used to create individual moods or reflect social moods. It is a flow of information, which is navigated by the mobile user on an everyday basis. This process can be confusing and difficult for one individual to follow. In this chapter it is argued that 'talk' is a source of alternative ways of social organization expressed in new ways of relating to the social networks to which individuals belong, expressed in criteria or strategies that individuals use in their everyday life to handle the confusion and overload of information in their lives. Thus, *being in the world* is associated with understanding in which context and *moods* individuals live their lives. Furthermore, deep and detailed analysis is required when individuals link or communicate with others because they *want to be with others*. Communication theory describes communication not only as the act of communication but also the understanding of the meaning in the communication (Habermas 1984a, 1984b).

The better the social network is known to the individual, the more the meaning of the communication is enhanced, giving communication more purpose. There are multiple interactions via networks and links of social order within work and home that add layers to the complexity of communication. One way to help with the analysis is to use social network theory to examine the models that could define the family, home and work, and then apply analytical research. This type of work is helpful but incomplete as it calls for more than social and communication theory to be integrated into the analysis, hence the need for a philosophical ground (*Dasein*) to understand the social medium.

In this chapter we take as a given the conceptualization through social network theory of the ideas that help understand the dynamics of ownership, attachment, use and attitudes towards mobile phones, which are widely owned by individuals. The explosion in the use of mobile devices in the last 15 years is well documented by researchers in the social field. As mobile devices and services become more ubiquitous elements in the world of the mobile individual, the need to adopt an individual and social culture towards those objects arises (Ling 1999).

Mobile technology is a factor in mixing and generating random social networks, based on the stronger social links already existing (e.g. family). This leads us to describe the social network of an individual on the basis of strong and weak social ties (Granovetter 1982), in which there is a level of stability in how members of the network establish relationships of adjacency with members of other networks through weak ties and meanwhile keep their strong ties in a close-knit cluster. Mobility, either physical or over time, challenges those strong ties and redefines the nature of the relationships we keep and value the most.

Mobile users need to keep their established social networks, but must also be able to join new ones which they would not otherwise have been able to tap into with conventional modes of communication, but which they can access via mobile devices. Some of those social networks have strong ties to the mobile user; others are weak or might recur over time without a particular pattern. To fully understand the structure of the interaction, a bridge is needed to reconcile the random idea of social contacts and the clustered reality depicted by the layers of media laid on individuals.

For mobile users, the ability to acquire the expertise required to tap into the new social networks created by the technology allows them to gain advantages over those who do not know or cannot access those networks. This apparent gain is an important integration resource when calling and joining networks aimed at social participation. The mobile user is driven by the need to tap into networks and the need to belong, to obtain relationships with others that are rich in narratives and informal in their practices, adding great social value to keep up with changes in the network. This 'social gain' can be enhanced or reduced depending on the strength of the ties created and the desire for participation attached to this activity (*the moods*).

Networks in general are a means to achieve goals that cannot otherwise be achieved on an individual basis, such as a political event. The need for *being in the world* produces a level of individual satisfaction that is highly psychologically rewarding (Myerson 2001). The more dense or closed a personal network is, the more satisfied the mobile user would be with the social aspects of it. Network members are relevant to provide instrumental help as well as validation of the social progress of the individual.

Mobile networks, then, are a favourable environment in which individuals can satisfy their eagerness to participate in the world (*being in the world*), whilst being with others according to their own particular situations (*moods*). An expression of these ideas is the use of the Internet as a source of information for news in the political sphere. This chapter analyses its role in the section below and links it to the ideas presented above.

The Internet, informational politics and mobile networks

In the previous section we have discussed in philosophical terms some of the factors that make mobile networks a very suitable environment to promote individual participation in social action. However, this is not the only medium that individuals can use to achieve this. Since its beginning, the Internet has been seen as a medium for expressing ideas of democracy and information politics to individuals and of the counter effects of the abuse of this medium, first within the research community and later on with the whole connected world, or Cyberspace.

The Internet is seen as an essential medium used by diverse civil and political organizations all over the world to promote awareness of their causes. Its prestige is as a privileged tool for acting, informing, recruiting, organizing, dominating and counter-dominating such causes. The virtual world of the Internet or Cyberspace has become a contested terrain (Castells 2001). It is interesting to note that *being in the world* is a concept extended to not only the real world but also the virtual world product of technology (Myerson 2001).

The Internet seems to have opened new avenues for social change that individuals have from its early stages adopted as their own. Individuals are motivated to participate in the Internet as a medium of communication by the need to express their own cultural values, as well as seeking to compensate for the gaps left by the crisis in their own vertical civil societies and the globalization of some causes such as women's rights, AIDS, etc. From political websites to personal weblogs, through discussion groups (public or by invitation only), on-line newspapers, mailing lists, etc., the Internet is a stream of political information that ontologically has many discourses at the core running simultaneously. Objectively, this ontology makes it difficult for one individual to grasp at a glance an overview of those core discourses.

In other words, if the Internet allowed the expression of individuals as civil society (historically the Internet was created in the United States and was protected by the Freedom of Speech Act, an argument that is constantly used to present the Internet as a tool for 'freedom'), this expression tool creates an excess of information available to the individual. But the overload of information does not match very well the postmodern citizen of our era, attached to the social networks we discussed in the previous section. Individuals struggle to cope with the overload of data (Sørensen 2002). Asian societies have experienced a process of openness due to the use of mobile technology, and share in common with users in the West the problems of overload and changes to the perception of identity and security.

For this reason, the success of the Internet as a situated social participation organization is more effective if related to other media, such as mobile networks. Further examples of how mobile networks can effectively organize social participation are also given later in this chapter. Such examples of social participation – *the being with others* – through mobile technology can be seen in the eyes of phenomenology analysis. To understand the individual world of the mobile user, it is necessary to feel the world, or, in other words, the individual must be able to differentiate between the *moods*, *being in the world* and *being with others* in each social situation. Here are some examples.

In 2004, a few US non-profits started collecting potential young voters' mobile numbers, planning to call and text-message them in the run-up to the election in November. The mood is felt here by the elections and the call to participation. The more reflective citizens of US society, in this case media, social researchers and modern thinkers, were alarmed by this practice called by those US nonprofit organizations. They expressed their concerns about the trivialization of democracy to a marketing survey (Longino 2004; Rushkoff 2004). The main aim of the call – as declared by the callers – was to try to engage young people in the voting process; the plan was to present the act of voting as something 'cool' to do, with the engagement of rock bands presented in TV programmes backed up by Internet websites and young people's social groups.

As pointed out by Rushkoff, trying to obtain young people's engagement in political activity in this way creates a problem: despite the aims, it tends to reduce the political process to a consumer choice. It could generate in young people a view that democracy turns into just one more consumer survey or focus group (Rushkoff 2004). The choice of candidate feels, for the target audience, more like a brand identity than a policy choice. The *moods* and the desire of *being with others* overtook the necessary reflection requested from individuals in a democratic process.

There are other cases where avoiding the trivialization of choice has had greater success. In mobile technology the main case is texting, or SMS (Short Message Service), which together with electoral politics are the new political partners of the twenty-first century. The use of SMS for political action is only in its infancy, but has already enabled citizens to topple governments and tip elections from Manila to Madrid, expressing the concept of *being in the world* as a reinforced concept of globalization.

Social networks associated with individuals give richness to these actions. They are a powerful engine for calls to action. Texting a simple SMS message to all the numbers listed in one mobile phone propagates a call for action very quickly. The electoral power of texting could be an early indicator of future social upheaval: whenever people gain the power to organize collective action on new scales, in new places, at different times, with groups they had not been able to organize before, societies and civilizations might change (Rheingold 2004).

Further cases in which mobile phones have been used as a political tool are the 1999 anti-WTO demonstrations in Seattle and the 2000 Manila 'People Power II'

(Agar 2004) demonstrations, which gave birth to the legend of 'Generation TXT' and signalled the end of the Estrada regime. Others are the election of President Roh in South Korea, the emergence of the Howard Dean candidacy in the USA, and the SMS-organized demonstrations in Madrid in the wake of the 11 March 2004 terrorist bombings and on the eve of the general election (Castells 2004). These successful uses of mobile phones to enhance social participation have been made possible by defining strategies for the calls for action.

Other examples that illustrate this global trend in the use of technology to keep up with democratic processes include the elections in Kenya and Ghana in December 2008, which were kept above board by monitors who used a network of mobile phones and radio stations, and also Twitter and other networking tools (Ghana Election 2008); India's Bharatiya Janata Party Congress Party, which uses SMS to maintain contact with the press and voters (Levin 2004); and in South Africa, SMS registration, which is part of the official voter registration process (Rheingold 2004).

The terrorist attacks in Spain occurred days before an election; in the first instance the Spanish Government declared ETA to be responsible for the attack. Rumours spread that links to Al-Qaeda had been found in the bombed trains. Officially, demonstrations were banned during the 24 hours prior to the election, but Spanish citizens used SMS to self-organize spontaneous demonstrations, as there was a general belief that the government was lying (Castells 2004). Texting traffic was 20 per cent higher than normal on the day before the election, and 40 per cent higher on Election Day. Castells, speaking at the LSE in 2004 (ibid.), said that texting was directly responsible for including two million extra votes in the elections that deposed Mr Aznar.

In all the examples above, a situation in the world (*being in the world*) together with individual or citizen participation (*being with others*) has produced an action to express the *mood* of the time. Heidegger's philosophical concepts are expressed through the use of mobile phones in the existing social networks, enhanced by the practices that mobile networks allow to their users. In the following section this chapter presents in more detail the study used for analysing the social framework of the Asian social context.

Organizing social participation in the context of Chinese society

The social context

The case to be presented below is a direct experience of the author of this chapter on the work done by members of the foreign community in Beijing during April 2003 to organize a demonstration in front of the US Embassy to express their individual opposition to the US invasion of Iraq.

The political situation in Iraq had been escalating over the weeks prior to the events to be described here. The United States, with some other countries (such as the United Kingdom), had accused the President of Iraq, Saddam Hussein, of holding Weapons of Mass Destruction and felt that this was a risk for the region, and using this argument decided to invade the country and send US troops to Iraq.

These actions were against the majority opinion of the United Nations Security Council. The media such as TV, radio, newspapers and the Internet were reporting in detail the actions of all parties involved.

The United Nations countries split in terms of their alliances into those that abstained from expressing an opinion about the US actions, those against it (such as France and Germany) and those supporting the United States (such as the United Kingdom, Poland and others). But for the countries that supported the United States, their government policies did not reflect the overwhelming opinion against the war that the citizens of those countries held. This was the case in the United Kingdom, Italy and Spain where public demonstrations against the war were organized in the streets of major capital cities. This is an example of a general mood of disagreement with the world.

Those demonstrations were organized by political organizations trying to reflect a more serious political agenda hidden behind the very controversial arguments used to verify the existence of Weapons of Mass Destruction in Iraq, and for their propagation used multiple types of media. The more established the organization, the more access it has to the media.

In that political context, there was a lot of 'talk' at all layers of society. In Beijing, where the author was staying for some weeks, we were able to observe first hand the amount of talk going on in the foreign community. Away from their own countries, these people were not allowed to express how they felt about the actions of their governments, especially the ones in favour of the invasion. It was at this time that those individual members of the Diplomatic Community in Beijing holding a different view of this matter to their own government (in cases where their government supported the invasion) decided to act as civil citizens (the right to express their opinion was held in high regard), and using their established social networks expressed, in public view, their disagreement with the invasion. Here the *being in the world* for an individual (that Heidegger defines as part of the *Dasein*) is a truth: individuals need to relate to the events in the external world to themselves to feel part of the world.

At all times the four organizers wanted to make clear that this was a citizen action not backed up by any institution, embassy or government. The Chinese Government had expressed their neutrality about – but not their agreement with – the invasion and was keen on keeping a political distance from any events that might express too much animosity towards the US actions. Hence the first objective of the four organizers was to define who would be interested in participating in a civil action. Most cell organizers lived or worked in the same area of Beijing (Cheoyang) and most of them had mobile phones and access to the web and e-mail.

The technology and the civil action

Most potential participants in the organization – at the time it was not clear that the political action would result in a public demonstration – used their mobile phones as the preferred medium to communicate their intentions to other acquaintances.

It was seen as a direct one-to-one media with immediate response (yes or no) to a call for participation.

At the early stage four people formed the mobile network associated with the organization of the demonstration. These individuals called or texted their closest associates inviting them to participate somehow and to bring ideas about the best way to express their discomfort with the situation in Iraq. There were many enthusiastic responses: messages in Chinese, French, English, Spanish, Italian, etc. were exchanged between diverse social groups. This is a social network in action, the *being with others* a means of expressing individual beliefs.

At this stage of early engagement, individuals propagated the idea of a formalized protest to their own networks of people on general terms, creating a *mood* for action. The feedback the self-nominated organizers had received from the social network was that either a printed manifesto in the media (Internet or newspapers) or a public demonstration would be ideal. The latter was chosen, as it would strategically have more impact in the conventional media such as TV. Some of the people initially contacted to participate in the organization of the protest were well linked to the International Press in Beijing.

Another suggestion from the people contacted as a way of reaching critical mass was to circulate an e-mail within a wider community of foreigners. For that purpose a Yahoo Open Forum was opened on the Internet and all the members already engaged in the group listed there the e-mail addresses of foreigners in Beijing that could be interested in participating. Further participation in the Forum was only possible by invitation. This was a special virtual space created in Cyberspace to exchange ideas.

Some of the organizers had chat facilities on their mobile phones and would keep simultaneous communications with more than two people at a time through this media to stimulate further participation. An invitation e-mail text was sent to around 250 people, primarily members of the many diplomatic services in Beijing. Soon the e-mail left the confines of the diplomatic world and was distributed to foreign companies and services. At some point there were 85 people actively engaged in the recruitment of people for the demonstration. All this engagement was low key as it was not clear when the demonstration was going to take place.

Wary of the delicate position of the host country, in this case China, the organizers approached the Chinese authorities for permission to demonstrate. There was a process of negotiation with the authorities, as the organizers were required to fully identify themselves, agreed to full responsibility for the demonstration action, and to define clearly the political scope of the demonstration.

The result

In compliance with the authorities, a tentative date was granted for the demonstration. A route that could be protected adequately by the police was agreed with the Chinese authorities. All members of the demonstration had to wear blue armbands and the leaders green arm bands and had to walk in rows of four. An emergency group met in one of the 'old *hutongs*' in Beijing and organized banners and armbands, plus slogans to be shouted out when walking.

Mobile phone calls and texting, e-mails and the Forum Group on-line again distributed all these demonstration instructions. It was highlighted to all participants that the success of the demonstration depended on the process agreed with the Chinese authorities. The demonstration meeting was scheduled for a Friday but it was suspended at last-minute notice due to logistics issues with the Chinese police. The demonstration went ahead the next day at 11 a.m.

From an organizational point of view it was a success. Plans existed to extend the demonstration for a support group to go and participate in the demonstration later in the day. To this scheduled demonstration the Chinese authorities did not allow the presence of foreigners. This demonstration did not take place as the organizers judged the agenda of this other demonstration to go beyond the Iraq War. In the foreign demonstration nearly 2,000 people participated: whilst marching, people were talking and sending texts to other people on their mobile phones to tell them how the demonstration was progressing. Also phone cameras were used to record the event. At the end of the one-hour planned march, the calls and texting continued. Observing around, it was possible to hear people talking to each other saying how good it was to be able to shout their own personal beliefs against the war. The demonstration was news in the international and local press and was praised for its order and clear political aim.

From the description of what happened when organizing the demonstration, we can deduce that there was some strategic use of the mobile networks to enhance political action. Listed below are some of the strategies identified in the case study.

Strategic use of the mobile networks in political action

The risks associated with the overload of information expressed in the previous section, together with some of the examples mentioned, allow us to define some primary ways mobile phones are used strategically to enhance political action.

Mobile phones as planning tools

Political strategists in organizations interested in a particular cause can use in the first instance their available contact list or members' databases (existing social networks with weak ties) of their supporters with mobile phone numbers and let the dynamics of networking at grassroots level take effect. In other words, because of this instrument, people who had not met before could contact each other and assist wherever they could in understanding the aims of the political case.

Mobile phones as one-to-one campaigning tools

Mobile phones enable the use of texting (SMS) on a one-to-one basis to stimulate active participation. The individuals contacted can then spread their views to the social networks surrounding themselves and exchange views with other similar social networks.

Mobile phones leading to action, political expression

The dissemination and extended political information allows coordination and effective political activism. Individuals can congregate and disperse very quickly as per the need of the political action demand.

However, like any powerful technology that amplifies collective action, mass-texting in politics has its pitfalls. Most obviously, the crowd and the mob have been manipulated by totalitarians in the twentieth century, using much cruder media technologies than are available today. Less obviously, texting is an unprecedented channel for misinformation and disinformation as well as an alternative channel for routing the truth around conventional media such as TV broadcasts (for instance, 150 million text messages propagated news about SARS in China at the early stages of development of the epidemic in a matter of minutes).

The case study presented in detail is a good example of how mobile phones can be used very successfully in organizing social participation in any social context. It seems to be that, through the merging and integration of the technology and social participation, individuals are able to express their unique entities as per Heidegger's *Dasein*.

Reflections and conclusions

Of all the arguments presented in this chapter, the sustainability of the philosophical concepts of *Dasein* in the postmodern world of mobile networks is the most remarkable highlight and of long-term reflective value for making sense of this world. The creation of a world in which the individualization of society seems to be a primary aim of the social forces is creating wider and looser social networks that can be drawn upon to act depending upon the context in which individuals live their lives. Those calls are not random but do have a concierge to channel effective ways of calling for social action.

This micro-cosmos of hypermobile citizens with access to multiple technical media devices to connect to the rest of the world has led to an overload of information in the daily lives of most people. The need for *being with others* makes individuals perceive as a 'must' the need to carry their mobile phones at all times, creating in fact an extra virtual space of communication and challenging our concepts of *being in the world*. If we can be in one place and talk to some people or be part of a demonstration, and at the same time by texting or calling we can *be* in touch with others, where are we? Are we in the physical space where our body lies and some of our actions occur, or in the virtual world of the mobile network?

Because of the demands of postmodern life, it is difficult for individuals to grasp an overview of social events or to engage in social action unless the loose network to which individuals belong is there to support them. In order for individuals to capture *the mood* of what is happening in the world, they have to increasingly rely on the advice of those mobile networks.

In this context, to be invited in an individual and not anonymous way by a member of the social network to which the individual belongs, and to be part of a social act such as a political demonstration, helps in the decision process each individual has to go through all the time when deciding what is personally relevant from the surrounding world. If a friend – who shares something in common with me – invites me to do something, it is a trustworthy personal act, compared to engaging on an activity that has been promoted by an impersonal leaflet handed over to me in the street.

If world societies are changing because of the changes in ideology and technology, the role of the individual as a citizen is also changing. Citizens are expressing political ideas through coordinated actions organized by mobile networks, which are highly effective and do not rely on media owned by governments or private companies.

In future it can be expected, as the technology advances, that more sophisticated networks of engagement will be discovered or created for the use of the mobile networks. How civil society is able to integrate this new social participation is still a challenge not addressed by most postmodern societies. So far we can count positive uses of the distribution of messages over mobile networks, which have had positive effects on the societies where they were deployed. We should also reflect on the negative ways this distribution of messages can be used and the effect it could have on the general well-being of the individual and the society.

Issues such as anonymity, security, monitoring and privacy were not discussed in this chapter but the author is aware of their importance in understanding the complete extension of the changes provided by mobile networks to the world we live in, as these issues do affect how individuals can be seen by the world and how we see the world.

Note

1 The author wishes to thank Ms Isolda Morillo for the opportunity offered to be a direct participant and observer in the 2003 demonstrations against the Iraq War. Her invaluable social contacts within the foreign diplomatic community made this research work possible. Thanks as well to France Telecom for the economic sponsorship provided to the author to visit Beijing to present the early version of this chapter in the conference Mobile Communication and Asian Modernities II in 2005.

References

Agar, J. (2004) *Constant Touch: a global history of the mobile phone*, London: Icon Books.

Bauman, Z. (2001) *The Individualized Society*, Oxford: Polity Press.

Berry, C., Martin, F. and Yue, A. (2003) *Mobile Cultures: new media in queer Asia*, Durham, NC: Duke University Press.

Castells, M. (2001) *The Internet Galaxy*, Oxford: Oxford University Press.

—— (2004) *Politics and Power in the Network Society*, LSE Miliband Public Lecture, London, 18 March.

Ghana Election (2008) http://ghanaelections2008.blogspot.com/2008_11_01_archive.html (last accessed March 2009).
Giddens, A. (1984) *The Constitution of Society*, Berkeley: University of California Press.
Granovetter, M. (1982) 'The strength of weak ties: a network theory revisited', in P. Marsden and N. Lin (eds) *Social Structure and Network Analysis*, Beverly Hills, CA: Sage.
Habermas, J. (1984a) *The Theory of Communicative Action I*, London: Heinemann.
—— (1984b) *The Theory of Communicative Action II*, Cambridge: Polity.
Heidegger, M. (1962) *Being and Time*, first English edition, Oxford: Blackwell Publishers.
Inwood, M. (1997) *Heidegger: a very short introduction*, Oxford: Oxford University Press.
Katz, J. and Aakhus, M. (2002) *Perpetual Contact: mobile communication, private talk, public performance*, Cambridge: Cambridge University Press.
Kleinrock, L. (1998) 'Nomadicity: anytime, anywhere in a disconnected world', *Mobile Networks and Applications*, 1: 351–57.
Levin, A. (2004) 'Chapter 14: Campaign Tools', *Extreme Democracy*. Online. Available: www.extremedemocracy.com/chapters/Chapter14-Levin.pdf (last accessed March 2009).
Ling, R. (1999) '"I am happiest by having the best": the adoption and rejection of mobile telephony', *R & D Report*, Vol. 15, Kjeller, Norway: Telenor.
Longino, C. (2004) 'Voter-registration groups call on mobiles', *The Feature Online*. Online. Available: www.thefeaturearchives.com/topic/Privacy/Voter-Registration_Groups_Call_On_Mobiles.html (last accessed March 2009).
Myerson, G. (2001) *Heidegger, Habermas, and Mobile Phones*, London: Icon Books.
Rheingold, H. (2002) *Smart Mobs*, Cambridge, MA: Perseus Publishing.
—— (2004) 'Political texting: SMS and elections', *The Feature Online*. Online. Available: www.thefeaturearchives.com/topic/Culture/Political_Texting__SMS_and_Elections.html (last accessed March 2009).
Rushkoff, D. (2004) 'Rock the (Wireless) vote.com', *The Feature Online*. Online. Available: www.thefeaturearchives.com/topic/Culture/Rock_the__Wireless__Votecom.html (last accessed March 2009) .
Sørensen, C. (2002) 'Digital nomads and mobile services', *Receiver – Vodaphone*, 6.

9 Social networks and individualism

Some issues on the role of the mobile phone in urban China

Boxu Yang

Introduction

Scholars in international or global communication emphasize the impact of media on social process or transformation. As far as developing countries are concerned, they often try to link media with national development regardless of their ideological orientation. That is, media, in particular the electronic ones, are regarded as agents having a powerful impact on social change. Nevertheless, such direct linkage may not always be clear or even exist. Evidence suggests that the impact of media on national development is much more complicated than previously proposed or theorized by either modernization theorists or dependency scholars in communication of the West, in most cases referring to the United States of America. This chapter tries to take a local approach and explore the mobile phone's social implications in urban China.

It is true that the modern media have been part of the social change or transformation since the beginning of industrialization, particularly in the West (Beniger 1986). However, the assessment of the role played by media in that change often ignores the power of two-way communication and media, while the impact of broadcasting media is exaggerated or misinterpreted. For instance, social changes often invoke a self-identity crisis. But how self-identity is related to the fixed phone or the mobile phone interaction is rarely addressed in the context of developing countries. Such relationships are significant because changes or reorganization in self-identity are closely associated with changes in social structure. That is, unless we have a good understanding of the identity change, we may have difficulty fully appreciating the dynamic of social transformation in the modernization and globalization process.

When Barry Wellman (2002) explains the social changes in the West, he uncovers three stages. They are door-to-door, place-to-place, and person-to-person (cf. Rheingold 2002). The shift from one type of space to another not only refers to changes in social structure but also illustrates the individuals' capacity as actors or agents. For instance, what has emerged from the place-to-place to person-to-person shift includes networked individualism (Wellman 2002). Therefore, in respect to the adoption of the mobile phone and China's modernization in the context of globalization, this chapter discusses two main questions: (1) Why does the mobile phone diffuse so rapidly? (2) If there is self-identity change, how is the mobile phone related to it? In order to answer the two questions, it may be enlightening to briefly review the identity concept as well as the past experience on the adoption and implication of the fixed phone in developing countries in general, and in China in particular.

Self-identity and social change

In an era of modernization and globalization, people are increasingly making contact with individuals who are remote both geographically and emotionally, for business or personal reasons, with the assistance of modern communication technologies. Those individuals may be friends, acquaintances, total strangers, or representatives of various organizations. When I take the initiative to contact someone, at least part of my self-consciousness is realized. When I am contacted by others, I become the object (the 'me'). But who am I in the social process? How does face-to-face interaction differ from fixed-phone- or mobile-phone-mediated communication and why does it differ in respect to self-identity?

Who am I? While such a question is very hard to answer, it is fundamental because the 'I':

1 is a key source of novelty in the social process;
2 is the location for our most important values;
3 constitutes something that we all seek – the realization of the self;
4 develops in an evolutionary process; that is, people in primary societies are dominated more by 'me', whereas in modern societies there is a greater component of 'I'.

(Mead 1934/1962)

That is to say, it is the 'I' which is related to the change. Thus, unless we have a basic understanding of the concept of 'I', we may have difficulty in fully appreciating the social change with which the two-way electronic media are associated. It is hard to answer because, although 'the "I" is the immediate response of an individual to others, we know the "I" only after the act has been carried out' (cf. Ritzer and Goodman 2004: 214). In fact, as discussed later, the diffusion of the fixed phone and mobile phone makes it even harder to answer.

To Mead, 'I' refers to a person's uniqueness, the impulsive, spontaneous, and reflective aspects (cf. Ritzer and Goodman 2004). These aspects can only be revealed through the realization of the self in social processes. In other words, we need to learn how the self-identity is constructed in order to know 'who I am'. But 'self' has two phases: the 'I' and the 'me'. 'The "me" is a conventional, habitual individual' (Mead 1934/1962: 197). It is through the 'me' that society dominates the individual. While the 'I' reacts against the 'me', 'the self is essentially a social process going on with these two distinguishable phases' (Mead 1934/1962: 178). In an evolutionary sense, the significance of such differentiation is that people in primitive societies are dominated more by 'me', whereas in modern societies there is a greater component of 'I'. Historically speaking, the greater figures have a larger proportion of 'I' than most others have. Even for the majority of ordinary people, anyone's 'I' may assert 'itself' and lead to changes in their day-to-day social situation. Here, the key point is that the specific exigencies of each person's life give them a unique mix of 'I' and 'me' (Mead 1934/1962; cf. Ritzer and Goodman 2004: 214–16).

Self can only be revealed through the discovery of the identity. 'Self-identity … is not something that is just given, as a result of the continuities of the individual's action-system, but something that has to be routinely created and sustained in the reflexive activities of the individual' (Giddens 1991: 52). Although the 'I' is not equal to self-identity, we may say that the 'I' permits the development of the self-identity at least to some degree. Just as the 'I' links to self-identity, the 'me' reflects the social identity. Social identities refer to the roles we play in society. Such roles include that of parent, neighbour, occupation, religious affiliation, gender, marriage status, and so on. That is, while self-identity enables us to see how one differs from others, social identity indicates how we are socially and structurally connected.

While some social theorists emphasize the structural impact on the role-related identity, others observe the influence of the 'I' on role identity (see Turner 2004: 369–83). McCall and Sommons (1978) indicate that role identity constitutes an 'imaginative view of oneself' in a position, often a rather idealized view of oneself. Thus, each role identity has a conventional portion linked to positions in social structure as well as an idiosyncratic portion constructed in people's imaginations (cf. Turner 2004). Nevertheless, role identities need to be verified or confirmed in interactions with others. When the verification fails, a person experiences a loss of efficacy and a greater sense of alienation, disaffection, and estrangement (Berke and Stets 1999; Stets and Tsushima 2001; cf. Turner 2004: 379). In such verification, the structure constraints are uncovered. Indeed, all structures impose limits and constraints on what individuals do when engaged in face-to-face interaction (Stryker 1980; cf. Turner 2004: 369). This leads to some, in my view, rather significant questions.

Sociologically speaking, industrialization or modernization and globalization all imply fundamental changes in social structures. To Mead, such changes may lead to a greater proportion of 'I'. Then, how are the fixed phone and mobile phone related to the expansion of the 'I' in terms of the construction of self-identity? According to the identity theory mentioned above, the structural constraints on individuals are mainly realized through face-to-face interactions. Thus, it becomes legitimate to ask what will happen to the role identity if the interaction becomes mediated, and to question how fixed phone communication is different from the mobile phone communication in respect to the social change.

The fixed phone and social change: from communities to social networks

Developed countries experienced industrialization. Telecommunication was part of the process. As the most pervasive medium, the fixed phone was the heart of the telecommunication system then because of its economic significance and social implications. Some of the most salient features of industrialization are the coordination of railway transportation, the fast flow of commodities, the construction of skyscrapers, and social mobility. These all demand the diffusion and application of the fixed phone. Nevertheless, the fixed phone is not on the list when the modernization theorists are asked to recommend electronic media to developing countries for their development. Instead, broadcasting media are regarded as the most effective tool for social change or economic development or both. After all, they believe that the first and foremost matter is changing the opinions, values, and attitudes of the masses for modernization to take place in developing countries. Broadcasting has a powerful impact on national culture. For instance, one of its most influential roles is arousing material desires and propagating consumerism. However, this is not consistent with the economic reality of developing countries.

Of course, we already knew very well that broadcasting was not an ideal tool for development by the early 1980s. Radio and television do an excellent job as an ideological tool, but are not necessarily related to the national development of developing countries in any meaningful sense; it is the diffusion and use of the fixed phone which is positively related to economic development in both industrialization and modernization processes. More importantly, along with a modern transportation system, the fixed phone has a powerful influence on social mobility and structuration as required by the market economy. Unfortunately, the indispensable role played by the fixed phone in modernization was ignored in a convenient way until the 1980s, and it is still overlooked to some degree. Instead, the broadcasting media were emphasized and introduced while the fixed phone was ignored for reasons such as that the nature of the broadcasting media is very good at assisting political or economic authorities or social elites to deliver their ideological messages to the audience. As an electronic network, the fixed phone has no such nature. That is, fixed phone communication has no audience. It is inherently interactive or democratic in nature in terms of the communication process. In other words, the fixed phone helps to contribute to the expansion of the 'I', while the broadcasting media assist to promote the 'me' in the self. Hence, it is necessary to readdress the significance of the fixed phone in order to assess the impact of the mobile phone. This section mainly elaborates on this particular point in respect to the recent mobile phone phenomenon.

The economic system throughout the First World's industrialization was the market economy. Now it has also become the dominant ideology for the modernization of developing countries. The market economy by and large demands and promotes the free flow or movement of goods, services, and labour. The precondition for the market economy to operate successfully is the free flow of necessary information. One of the key media for the information flow is the fixed phone. That is to say, the fixed phone is diffused not for community members to chat with each other but for (1) members of a business sector to exchange commercial information, and for (2) members of a social network to contact each other from a distance. Under the market economy, community cannot exist for it contradicts the concept of social mobility which is crucial to business expansion.

The new social structure which emerged during industrialization is the social network. When we start to enjoy the convenience of a fixed phone for social purposes, we already live our lives in social networks.

When I refer to the fixed phone's role in social change in industrialization or modernization, I do not suggest that it determines the structural transformation. Rather, what I mean is its matching with the social structure in industrial society. Such society is, of course, based upon social networks rather than communities. Social networking has no respect for physical boundaries to a large degree due to the assistance of the modern transportation system (namely, highways, cars, etc.) and telecommunication networks. This often means an individual's detachment from the local communities and the enabling of social network formation or reproduction. The coupling occurs when fixed phone networks connect social networks (Wellman and Tindall 1993). The result is structural transformation which is part of the industrialization or modernization process.

The structural change or transformation indicates the transition from community life to social network life. Life in a social network involves much more personal freedom due to the instability of family and the disappearance of neighbourhood-oriented community. That is, our social life is less and less monitored and sanctioned by our neighbours because our social relations go beyond physical boundaries (Haythornthwaite and Wellman 2002). When we speak to someone on a fixed phone, we only deliver our voice messages. The rest (such as facial expressions, body movements, appearance, social distance, and the like) is missed partially or entirely. But non-verbal communication is often crucial for social identities. This implies the expansion of the 'I' and fewer constraints in terms of the creation of self-identity. It is true that fixed phone communications are positively related to face-to-face meetings (Wellman and Tindall 1993). Nevertheless, such meetings are rarely casual. Rather, not only are they prepared both psychologically and physically, but also fewer role identities may be involved or revealed in the interaction due to the fixed phone. The fixed phone simply makes it easier for an individual to control or manipulate the impression given to others when they appear in the presence of others (cf. Goffman 2004: 41–46).

But it is important to recognize the fact that the fixed phone is mostly located in private spaces. They differ from the public ones because of their exclusiveness. However, the exclusiveness is related to power. For instance, an employee may find it difficult to exclude his superior when he is making a business call. Similarly, in most cases, a husband has no right to ask his wife to leave the room when he is talking to someone on the phone at home. Thus, the fixed-phone-related space is not a personal space either. That is, fixed phone communication is conditioned by the private space because it can be monitored and interfered with. The sanction not only suggests how the structural constraints exist but also reveal the importance of social identities or role identities. Indeed, it sets the technological limits to the development of 'I'.

The social and media experience in China's modernization was dramatically different from that in industrialization in the West not only because China preferred broadcasting media to the fixed phone in terms of media adoption before the 1990s, as most of the developing countries did, but also because the traditional Chinese community was very much a blood-related clan. The traditional Chinese community or clan is organized in terms of family ties and lived within geographical constraints, mainly in the form of a village. That is, a clan is formed on the basis of a shared ancestor. Thus, it is not uncommon to find that the members of a clan all have the same surname. The extension of one's social ties beyond the clan often takes the form of marriage arranged by the elders of both parties involved. Of course, the social interaction is face-to-face in nature and guided by the necessary rituals. Accordingly, the 'I' shrinks and the self-identity is more or less frozen in the process of structuration. It is the 'me' or role identity assigned that really matters. This particular character seems somehow consistent with the social life of a Western community.

Of course, even in traditional Chinese society, the social ties may include non-family individuals to some degree. For instance, such ties often existed among the elites or ruling-class members in various dynasties in history. However, this kind of tie is often established and maintained for the survival and prosperity of the family or household. This was particularly true after the Han dynasty and continued for some 2,000 years (cf. Yan 2005). Individuals who live in the blood-related clan and who wish to live comfortably have to fully identify themselves with the community by constructing and revealing their self-identities through the various rigid roles they play. For instance, we receive and give emotional and material support to each other to show we are caring members of a community because our life starts and ends in the community to a large degree. The interaction is based upon face-to-face communication. There are not many personal secrets or privacy issues in regard to social relations among the neighbours at least because the community members are able to watch each other constantly. In order to survive in such a closely tied and highly structured life, people often have to hide their true emotions (such as smiling or laughing) and other behaviours (like hugging and kissing) in face-to-face interaction. In other words, there is little room for the 'I' to grow for the vast majority of the clan members. However, clan life as such is inconsistent with industrialization or modernization, which demands social mobility. So the clan has been challenged and has been disintegrating in China. The identities are redefined and reorganized at the same time. This is one of the main social factors that contribute to the emerging Chinese version of 'Risk Society' (cf. Beck 1992).

I have argued that industrialization or modernization demands social mobility. What is meant is that there has to be a pool of labour readily available to the corporations so that they can organize production or conduct sales at any time and place. While the labour required has to be obedient and willing to be relocated, the labour force is also given the liberty or self-consciousness to freely take action under the terms of the market approach. This means that the emotional and material aid from the geographically bounded community or clan gradually becomes dispensable. As far as social welfare is concerned, things that used to be taken care of by the community and clan now are gradually covered by various organizations or agencies. Thus, emotional support becomes one of the major motives and practices for social networking. Individuals start to have the choice to interact with each other and form ties more and more along the lines of interests and personality (cf. Aronson 1977). This is very significant in the Chinese context.

To today's urban Chinese, life in a social network means having dinner with friends in restaurants, singing with friends in karaoke bars, and seeking resources from friends for personal gain. Those friends are non-geographically bounded network members. They become friends not only because they are alike to some extent but also because they allow each other to show 'who I am'. That is, there is more space for the novelty aspect and realization of the self in the social process. More specifically, when those kinds of activities are organized through the fixed phone, they signal that not only are structural constraints filtered to some degree but also the proportion of 'I' in the self is increased in the process. However, a bigger component of the 'I' does not necessarily mean a better realization of self or self-identity. In fact, it may invoke an identity crisis under the conditions of post-industrial society. This is crystallized in the case of the mobile phone.

The mobile phone and social change: from networks to individuals

While the fixed phone is associated with social networks or networked society, the mobile phone is related to what Barry Wellman called 'networked individualism' as far as social structure is concerned. Haythornthwaite and Wellman describe the networked individual thus:

> Rather than relating to one group, they cycle through interactions with a variety of others, at work or in the community. Their work and community networks are diffuse and sparsely knit, with vague, overlapping, social and spatial boundaries . . . People now go through the day, week, and month in a variety of narrowly defined relationships with changing sets of network members . . . The security and social control of all-encompassing communities had given way to the opportunity and vulnerability of networked individualism.
>
> (Haythornthwaite and Wellman 2002: 33)

The context for such a shift is globalization. One of the preconditions for globalization to occur is the triumph of the free-market economy worldwide. The globalization process has demonstrated the power of transnational corporations. The theory of three worlds has clearly become out of date in the process. More or less, we now live in a world consisting of nation states, non-governmental organizations, and transnational corporations. In other words, transnational corporations have become one of the three major players on the world stage. Their activities are facilitated by and organized through the Internet and the mobile phone. While the Internet signals the transition from international communication to global communication, the mobile phone is more related to local communication in the sense of glocalization. Regardless of their differences, it is safe to say that the two media are liberating tools in terms of social networking.

Rogers (2000: 84) observes that '*Globalization* is the degree to which the same set of economic rules applies everywhere in an increasingly interdependent world'. The significance of globalization is that economic rules not only apply everywhere but also affect almost every aspect of our lives. When the transnational corporations and non-governmental organizations reach the locality where the social networks are situated, the networked society starts the process of glocalization. Networked individualism emerges from the process of glocalization with the assistance of the Internet and mobile phone. To the urban Chinese, this may refer to social liberation without freedom to some extent because the liberation creates opportunities as well as vulnerabilities.

When the fixed phone was diffused during the process of industrialization, most individuals 'lost' their neighbourhoods but gained friends city-wide in the form of a social network. For instance, some 30 years ago, it was found out that the great majority of East Yorkers' intimates live within Metropolitan Toronto (Wellman 1979: 1021). Now it seems that the message delivered by the mobile phone tells us that it may be possible to restructure our lives once again. The fixed phone differs from the mobile phone because the former is mostly located in a private communication item, while the latter is a personal communication item. Accordingly, although the fixed phone provides a certain degree of freedom for reaching out, it does not necessarily discourage social identities because either colleagues or family members may be present during the fixed phone communication. On the contrary, the mobile phone is not location specific. Thus, it enables the creation of personal spaces that allow one to maintain a sense of privacy and to control the amount of stimulation one receives during interaction with another person (cf. Dosey and Meisels 1969). Although the conception of personal space is highly variable, it can be determined on a habitat level by profession, livelihood, and occupation. Moreover, it is important to acknowledge that personal space can also be heavily affected by a person's position in society, with a more affluent person demanding a larger personal space. A more affluent person is more likely to have a greater proportion of 'I'. The point is that every mobile phone user is able to create the personal space desired, although it is not a rigid physical space. In fact, most of the time they are able to create their own personal spaces anywhere they wish in order to communicate using the mobile phone (see Katz 2006). Thus, it becomes reasonable to assume that mobile phone users have the potential to increase the proportion of 'I' because of their ability to manipulate the space.

The mobile phone is inherently personal in nature because it simply has very few time–space constraints. People can change sets of network members in daily social interaction because the mobile phones or laptops people carry with them have no fixed locations. They are somewhere but nowhere. When one interacts with different sets of network members, one often needs to change one's self-identity and social identity. The frequent identity switch not only implies exposure of different dimensions of 'I' and 'me', but also means different space requirements. For instance, when one switches his role from a manager to a lover, he very likely needs to change the space from a private one into a personal one. The mobile phone is different from the fixed phone precisely because it allows such role switches and space changes to occur. This means liberation to a much greater degree for the development of self-identity or the 'I'.

When the mobile phone is diffused to such a remarkable degree in China, we have reason to believe that the individualism Wellman described is growing. In urban China, this individualism often invokes social identity renegotiation. In the social networks which emerged and developed during industrialization and post-industrialization, people managed their lives through playing a set of relatively stable roles with a consciousness of the identity hierarchy of prominence, which is mainly based upon the local situation. Under the context of modernization and globalization, the urban Chinese have to encounter both the local tradition and the foreign influence in an intense way. Recent research findings have revealed that mobile phones enable migrant workers to not only survive their identity uncertainties but also maintain their traditional social ties and develop new ones in China. Those migrant workers are from rural China. Their frequent mobile communication and obsession with the mobile phone per se are not only good illustrations of the intensive conflicts among various social forces but also persuasive demonstrations of the ingenuity of agents in terms of constructing and reconstructing their identities and social networks (Chu and Yang 2006; Law and Peng 2006).

The local habits or traditions are passed from history to the current day through social reproduction. The original teachings of Confucius offered a hierarchical social structure to the Chinese more than 2,000 years ago. In his teachings, although a superior (such as the ruled) has tremendous power over his subordinates or subjects, he has to earn respect, loyalty, and enthusiasm from them by treating them with dignity and kindness (Confucius 1993: 45, 46). Similarly, Confucius emphasizes that 'the father should act fatherly so that the son can follow in a manner most appropriate to his self-identification. The son's filiality is conceived as a response to the father's kindness' (Tu 1985: 124). However, a father should not or cannot be checked if he fails to set an example for his son as a loving and respectable person in terms of Confucianism. In fact, the son should still continue to be respectful and not show disobedience even if his father does not act in a fatherly way. This is important because this kind of social relation is vital for nurturing the desired vertical relationships between the ruler and subjects.

For relationships among friends, Confucius warns his followers, 'Do not make friends of those who are not up to your own standard' (Confucius 1993: 34). Such a warning implies that friends are beneficial only when they are honest (trustworthy), sincere, or knowledgeable. Taking his teachings as a whole on relationships, one needs '[c]ourtesy in private life [be respectful at home], reverence in handling business [serious at work], loyalty in relationships with others [faithful in human relations]' (Confucius 1993: 51–52). Nevertheless, those teachings have been gradually transformed into self-serving political doctrines for the ruling class in social reproductions during the past dynasties. The key principles for the doctrine are the three bonds (*San Gang*). That is, 'the ruler, the father, and the husband are to be the standards of the ruled, the son, and the wife' (Chan 1973: 272). In this way, the political interests of the state and the welfare of the family are intertwined in a rigid hierarchical structure. This is the base for our understanding of the Chinese social networks and identities which emerged in the modernization process.

Evidence about the social structure in overseas Chinese communities is well documented. While the family remains the basic survival unit, under the contemporary and different political systems of the host countries, it is somewhat divorced from the State politically. However, the research findings indicate that the hierarchical ranking system of father–son, husband–wife, elder brother–younger brother, and boss–subordinate is still the main structure. The essence of such a ranking system is control (see Lim 1983; Omohundro 1981, 1983; Oxfeld 1993; Redding 1990; Redding and Whitley 1990). As far as relational properties are concerned, 'you trust your family absolutely, your friends and acquaintances to the degree that mutual dependence has been established and face invested in them' (Redding 1990: 66). In contrast to the high degree of trust among the family members, there are very few institutionalized trust relations beyond kinship groups in Chinese society. The commitment beyond kinship groups is contingent and can be withdrawn (Redding 1990; Tam 1990). This is the social context for the understanding of self-identity and social identity in the Chinese modernization process. It is also the starting point for the realization of self in glocalization.

Not only do the family and state become less integrated, but the kinship group or family network also appears to be less close-knit during Mainland China's modernization process. Nevertheless, families remain in stable locations and can be identified in a relatively easy way. What the fixed phone has done is to facilitate the social change. But when transnational corporations become part of China's modernization efforts, glocalization occurs and shows fundamental structural changes in social relations. The former indicates the disassociation between households or family networks and government positions to a large degree, while the latter shows the nature of networked individualism. That is, the relational properties such as trust and loyalty are limited only to immediate family members such as parents and children. People constantly switch social networks in their daily lives. However, with little trust and loyalty left, the networked members are transformed into symbols for each other, and the interpersonal communication between or among the members is full of uncertainty. Accordingly, social interactions become a risky business. Thus, constant and mutual identity verifications are not only necessary but also vital to the existence of social relations. Failure to verify identity may lead to a loss of efficacy and greater sense of alienation, disaffection, and estrangement (cf. Turner 2004: 279–380). In this way, we may understand the rapid diffusion of mobile phones and the huge number of short messages (SMS) people send to each other on a daily basis.

As far as urban China is concerned, it seems that it is quite easy to reach out and connect to somebody by means of the mobile phone. Indeed, the sheer volume of mobile phone communication appears to show that the Chinese have been doing just that. However, due to the lack of trust and loyalty, especially when the communication is motivated by self-serving reasons, social identity becomes even more important because it is often regarded as an object for exploitation. Thus, some Chinese sociologists have shown deep concern at the seemingly common phenomenon of killing the relationships among friends or acquaintances for personal gains by cheating in current Mainland Chinese society (see Sun 2002; Zheng 2003). Of course, the interaction between non-networked members may involve various bribing activities (sometimes termed a gift) for personal gain. In both cases mentioned above, the mobile phone seems particularly appropriate for the communication demanded in such interactions.

The mobile phone is ideal for communication because it is personal in nature. Its mobile feature not only allows for the anonymity of both parties involved but also offers the opportunity to take advantage of others. Moreover, because the text- or voice-based mobile phone greatly reduces the richness of non-verbal communication, it particularly fits the rational interactions that focus on exchange in today's marketing age. More importantly, people now go through the day, week, and month in a variety of narrowly defined relationships with changing sets of network members. The narrower the relationship is defined, the fewer the social identities revealed in the communication. The higher the frequency of changing sets, the higher the volume of text- or voice-based mobile communication. Accordingly, one's social identities involved in the interactions become differentiated and fragmented. Without a full grasp of one's social identities, it is difficult to nurture relational properties such as trust and loyalty. Thus, vulnerability becomes part of the networked individualism in urban China.

When social identities are filtered through the mobile communication to some degree, self-identities become more prominent. Indeed, the differentiation and fragmentation of social identities leave the space for the 'I' to expand. But it is the social identity that allows us to feel that we are part of a group or organization or society. Hence, the expanded 'I' may permit the development of multiple and active self-identities in order to face the situation in which individuals frequently switch and interact with different sets of network members in order to reduce alienation, disaffection, and estrangement. The point is that the urban Chinese face particular challenges in a world of differentiation and fragmentation of social identity because of the lack of trust and loyalty even between or among 'friends' or network members. One has to be constantly alert and frequently change 'faces' (identity) in daily interactions with different network members in order to be part of something. Failure to do so simply means isolation. How important is the mobile phone to the development of self-identity and associated interaction? A young, white-collar urban Chinese recalls her working day:

> I was sitting in the office and working on the papers that had piled up on my desk [during the whole morning]. When time for a break [at noon], I felt that there was something wrong. It was simply not normal. I scratched my head and paused for a few seconds. Gee, my mobile phone did not ring today…
>
> For the whole afternoon, the tension was building up between me and a quiet mobile phone in the office. I was very alert and looking forward to hearing a ring. I started to panic – I was forgotten today. My lovely mobile phone innocently stared at me, a sad woman who was completely captured by modernization …

> My way home seemed long and lonely [after work]. There were so many cars and people [on the street], but I felt that I was just a grass seed that was thrown out to an ocean. In an age of modern communication technologies, we could contact anyone in anytime as we wish. However, the [social] distance becomes so clear and remote …
>
> (Wang 2003)

The differentiation and fragmentation of social identities demand the 'I' to expand. In so doing, the self-identities realized or created have to have the spirit of networked individualism or dynamics of reaching out. Otherwise, there would be no social relationships although the mobile phone is all about social interactions. Indeed, it seems that more and more Chinese experience the hardship of daily interaction lacking trust and loyalty in an age of Chinese networked individualism.

Discussion and concluding remarks

China is in a unique situation when it comes to the adoption and post-adoption of the mobile phone. It is unique because of its social tradition and current involvement in modernization and globalization. The fixed phone is diffused in China for it is indispensable to the modernization process. The mobile phone is welcomed because of its association with globalization. The difference between the two media reflects the difference between modernization and globalization. That is, the former is always stationed in a fixed location while the latter is mobile, which suggests a shift from private space to personal space. The space shift is consistent with the shift from a place-to-place to a person-to-person community (Wellman 2002). This pushes what Giddens termed the time–space distanciation to a new level (Giddens 1991). Indeed, modernization is more a national endeavour, while globalization often refers to transnational activities.

But both modernization and globalization pose severe challenges to the traditional Chinese social structures. At the same time, social reproduction shows the impact of Chinese tradition on the contemporary social relations. The mobile phone phenomenon may be regarded as an indication of this complicated process of glocalization in China. Thus, the fixed phone becomes available to the public when their clan-oriented community begins to disintegrate. However, in the same way that fixed phones are stationed, social ties are still very much location specific even though the ties are beyond the neighbourhood. Relatively stable identities of individuals are realized and created through the interaction among networked members. Stable locations impose certain constraints on attempts to breach the trust in social relations because most of the interpersonal communication occurs in private spaces. In other words, although the neighbourhoods have disappeared in the process of modernization, social networks are constructed in the realization of 'I' under the constraints of private space to some degree.

The rapid diffusion and implementation of the mobile phone in the age of globalization has touched the nerve of the traditional Chinese social structures. The

Chinese community is not necessarily a community in the Western sense. Regardless of Confucius and his followers' teachings, its hierarchical and household nature suggests that it neither allows personal space nor has the necessary relational resources to deal with a social world in which the reference point has no fixed location. The lack of trust and loyalty among network members implies remote social distance in urban China. For better or worse, the mobile phone becomes the medium employed to bridge the distance. Indeed, the process of glocalization does demonstrate the networked individualism. However, the individualism in the Chinese context often implies killing social relationships between or among friends or relatives for instant personal gain at the current stage of glocalization.

In the process of glocalization, the 'I' experiences the time–space distanciation in a much more elaborative way because the mobile phone enables the creation of personal space. This gives the 'I' the unprecedented freedom to expand. The expansion of the 'I' permits the creation and realization of fluid and multi-self identities in order to face the ever-changing daily interactions. However, the development of those 'I' characteristics are not only conditioned by the traditional Chinese social structures but also guided by basic instincts which may be more easily unleashed as consumerism under globalization. Thus, we may say that the fluid and multi-self identities permitted by the 'I' are more likely biased towards the materialistic dimension of consumption.

When a society is increasingly organized around consumption, it is reasonable to argue that the roles individuals play become more and more materialistic in nature. The point is, 'by virtue of playing a role, individuals incorporate meanings and expectations associated with this role into their identity in the situation' (cf. Turner 2004: 376). If the meanings and expectations incorporated into the self-identities are commodity fetishism oriented, the fluid and multi-self identities will be 'one-dimensional' in essence. Accordingly, a social identity is valued only when it is associated with scarce resources. Thus, networked individualism-oriented interaction becomes a calculated and rational exchange. The urban Chinese experience discussed above may be the extreme version of such individualistic interaction. That is, the property of belongingness in social identity has disappeared.

The Chinese badly need love and affection because social relations have been killed for personal gain to some degree in the process of glocalization in urban China. This may be why mobile phone communication is so pervasive there. Indeed, people become free from space or location to a great extent and interact with a variety of others in their daily routines. But in so doing, the urban Chinese may experience great isolation and anxiety because of the lack of trust and loyalty. We send and receive mobile phone signals to verify the existence of ourselves and others for we have so many compromised identities and masked faces. This invokes identity uncertainties. Nevertheless, it is crucial to recognize the fact that the Chinese had never had the personal space to realize their 'I' until the coming of mobile phones. It is also true that the Chinese had never enjoyed the public space until the diffusion of the Internet in China. What we know now is that the Chinese users are very active on those platforms. They are not only reaching out on an interpersonal level through the mobile phone but also taking the role of participant in various discussions on the Internet platforms. Moreover, the Internet communication is increasingly mobile due to the technological advances. While it is hard to predict what will eventually emerge from the mobile platforms in respect to the social relationship for the Chinese, it is safe to say that the self-identity will be more richly developed as long as the role of participant is realized on the Internet. Otherwise, the vast volume of mobile phone communication may only suggest social 'obesity', not social relationships in China.

References

Aronson, S. (1977) 'Bell's electrical toy: what's the use? The sociology of early telephone usage', in I. de Sola Pool (ed.) *The Social Impact of the Telephone*, Cambridge, MA: MIT Press.

Beck, U. (1992) *Risk Society: towards a new modernity*, trans. M. Ritter, London: Sage.

Beniger, J. (1986) *Control Revolution: technological and economic origin of the information society*, Cambridge, MA: Harvard University Press.

Berke, P. and Stets, J. (1999) 'Trust and commitment through self verification', *Social Psychology Quarterly*, 62: 347–66.

Chan, W. (trans.) (1973) *A Sourcebook in Chinese Philosophy*, Princeton, NJ: Princeton University Press.

Chu, W.C.R. and Yang, S. (2006) 'Mobile phones and new migrant workers in a south China village: an initial analysis of the interplay between the "social" and "technological"', in P.L. Law, L. Fortunati and S. Yang (eds) *New Technologies in Global Societies*, Hackensack, NJ: World Scientific.

Confucius (1993) *Analects*, trans. R. Dawson, Oxford: Oxford University Press.

Dosey, M. and Meisels, M. (1969) 'Personal space and self-protection', *Journal of Personality and Social Psychology*, 11: 93–97.

Giddens, A. (1991) *Modernity and Self-Identity: self and society in the late Modern Age,* Stanford, CA: Stanford University Press.

Goffman, E. (2004) 'The presentation of self', in E. Howard (ed.) *Classical Readings in Sociology*, Beijing: Peking University Press.

Haythornthwaite, C. and Wellman, B. (2002) 'The Internet and everyday life: an introduction', in B. Wellman and C. Haythornthwaite (eds) *The Internet in Everyday Life*, Oxford: Blackwell Publishing.

Katz, J. (2006) *Magic in the Air: mobile communication and the transformation of social life*, New Brunswick, NJ: Transaction Publishers.

Law, P.L. and Peng, Y. (2006) 'The use of mobile phones among migrant workers in southern China', in P.L. Law, L. Fortunati and S. Yang (eds) *New Technologies in Global Societies*, Hackensack, NJ: World Scientific.

Lim, L. (1983) 'Chinese business, multinationals and the states: manufacturing for export in Malaysia and Singapore', in L. Lim and L. Gosling (eds) *The Chinese in Southeast Asia, Vol. 1*, Singapore: Maruzen Asia.

McCall, G. and Sommons, J. (1978) *Identities and Interactions*, New York: Basic Books.

Mead, G. (1934/1962) *Mind, Self, and Society: from the standpoint of a social behaviorist*, Chicago: University of Chicago Press.

Omohundro, J. (1981) *Chinese Merchant Families in Iloilo: commerce and kin in a central Philippine city*, Quenzon City, Metro Manila: Ateneo de Manila University Press.

—— (1983) 'Social networks and business for the Philippine Chinese', in L. Lim and L. Gosling (eds) *The Chinese in Southeast Asia, Vol. 1*, Singapore: Maruzen Asia.
Oxfeld, E. (1993) *Blood, Sweat, and Mahjong*, Ithaca, NY: Cornell University Press.
Redding, S. (1990) *The Spirit of Chinese Capitalism*, Berlin: Walter de Gruyter & Co.
Redding, S. and Whitley, R. (1990) 'Beyond bureaucracy: towards a comparative analysis of forms of economic resource co-ordination and control', in S.R. Clegg and S. Redding (eds) *Capitalism in Contrasting Cultures*, Berlin: Walter de Gruyter & Co.
Rheingold, H. (2002) *Smart Mobs: the next social revolution*, Cambridge, MA: Perseus Books.
Ritzer, G. and Goodman, D. (2004) *Modern Sociological Theory*, Beijing: Peking University Press.
Rogers, E. (2000) 'Informatization, globalization, and privatization in the New Millennium', *Asian Journal of Communication*, 10: 71–92.
Stets, J. and Tsushima, T. (2001) 'Negative emotion and coping responses within identity control theory', *Social Psychology Quarterly*, 64: 283–95.
Stryker, S. (1980) *Symbolic Interactionism: a structural version*, Menlo Park, CA: Benjamin/Cummings.
Sun, L. (2002) *Cong 'Shashu' dao 'Lingruo'* [From 'killing the acquaintance' to 'bully the weak']. No longer available online: www.eobserver.com.cn/ReadNews.asp?NewsID=1023 (accessed 8 May 2002).
Tam, S. (1990) 'Centrifugal versus centripetal growth processes: contrasting ideal types for conceptualizing the developmental patterns of Chinese and Japanese firms', in S.R. Clegg and S. Redding (eds) *Capitalism in Contrasting Cultures*, Berlin: Walter de Gruyter & Co.
Tu, W.M. (1985) *Confucian Thought: selfhood as creative transformation*, Albany, NY: State University of New York Press.
Turner, J. (2004) *The Structure of Sociological Theory*, Beijing: Peking University Press.
Wang, H. (2003) 'Jintian Shouji Meiyou Xiang' [There is no mobile phone communication today], *Nanfang Zhoumo* [The Nanfang Weekend], 26 February, B2.
Wellman, B. (1979) 'The community question: the intimate networks of east Yorkers', *American Journal of Sociology*, 84: 1021–31.
—— (2002) 'Little boxes, glocalization, and networked individualism', in M. Tanabe, P. Besselaar and T. Ishida (eds) *Digital Cities II: Computational and Sociological Approaches*, Berlin: Springer-Verlag.
Wellman, B. and Tindall, D. (1993) 'How telephone networks connect social networks', in W. Richards, Jr and G. Barnett (eds) *Progress in Communication Sciences, Vol. XII*, Norwood, NJ: Ablex Publishing Corporation.
Yan, A. (2005) *Han Jin Jiazu Yanjiu* [Studies on the Households of Han and Jin Dynasties], Shanghai: Shanghai Renmin Chubanshe [Shanghai People's Press].
Zheng, Y. (2003) 'Zou Xiang Shashou Zhilu' [To the road of killing acquaintances], in Y. Zheng and S. Peng (eds) *Zhongguo Shehui Zhong De Xinren* [Trust in Chinese Society], Beijing: Zhongguo Chengshi Chubanshe [China City Press].

10 The identity and sociability of the mobile phone in China[1]

Leopoldina Fortunati and Shanhua Yang

Introduction

What is the 'identity' of the mobile phone in China? What are the processes related to its social construction? Are they different from those recognized in Western countries? If so, how different? Before answering these questions, it is necessary to explain why we use the term 'identity' in discussion of the mobile phone. We are encouraged to do so, first of all, because of the notable tendency for people to attach to an object (mobile phone) certain metaphors and characteristics which were once only applicable to human beings. The tendency is manifest in expressions such as 'machines that become us' (Katz 2003), 'when things start to think' (Gershenfeld 1999), and 'intellective machines' (Maldonado 1997). This happens because technological objects, especially those related to information and communication technologies (ICTs), are increasingly humanized (Contarello *et al.* 2008). The ontological distance between technological objects and human beings is becoming shorter (Latour 1998) and, in particular, the mobile phone is subjected to strong processes of hybridization with the human body (Fortunati *et al.* 2003).

Second, there is a general consensus that the identity of human beings is built not only by means of belonging to different social groups such as gender and generation, but also by means of the objects people own, show and use. So, the object is a generator of one's identity (Douglas and Isherwood 1979). Not by chance did William James talk about 'material identity' (1890). Being strongly entailed in the identification processes of human beings, the object necessarily assumes a reflexive identity, a secondary identity. This secondary identity of the object is further strengthened by another modern characteristic of the object itself, which is its being part of a system of information and signs (Douglas and Isherwood 1979). Being a receptor and transmitter of information and signs, the object allows the sharing of content, social meaning, and symbolic value (Eco 1968). All of these contribute to reinforce the identitary structure of the object.

Third, we discuss the mobile phone in terms of its identity because the technological object is constantly read, defined, and built by its users (Latour and Woolgar 1979). More precisely, the identity of the mobile phone springs from the relationship between producers, distributors, and advertisers on the one hand, and buyers, users, consumers, and costumers on the other. It also reflects how the mobile phone is related to the world of objects at large. This is a process of co-construction of the technological object, as social and cultural artefact, in its first stage of 'domestication', as it was termed by Silverstone *et al.* (1992). In particular, in the European project SIGIS, we found that the domestication process of the mobile phone had a particular rhythm in the sense that it was the users in the second and third waves of adoption who were the more proactive in defining and constructing 'what the mobile phone is' (Fortunati and Manganelli 2003). This domestication process has also been studied at the psycho-sociological level, in monitoring how the mobile phone has been metabolized in the system of social thought and the way in which it has been integrated conceptually (Fortunati and Manganelli 2008).

It is in this sense that we are motivated to examine the identity of the mobile phone as a centrepiece of our inquiry. For the same reason, we find the 'objective' definition of the mobile phone, as well as of all other ICT devices, largely unsatisfactory due to its inability to expose the increasing social and cultural complexity of this device. Such complexity reflects not only the enormous immaterial efforts embodied in the object (Baudrillard 1968, 1972) but also the multi-layered use and consumption processes therein.

So, again, what is the 'identity' of the mobile phone in China? What are the processes related to its social construction in the Chinese context? Are they different from those recognized in Western countries? If yes, how different is it? In Western countries the identity of the mobile phone has been investigated in many research projects, which have registered for instance the shift of the mobile phone from being a mobile device to being a personal device, from being an oral medium to becoming also a written medium, and so on. These studies underline many aspects, especially those related to the domestication, integration, and counter design of this device produced by users (Haddon *et al.* 2005). The book *Thumb Culture: The Meaning of Mobile Phones for Society* (Glotz *et al.* 2005) offers an exhaustive list of user interventions on the mobile phone as technological artefact. The map of the design variants reported in this book begins from the *individualization* of the mobile phone, its *sedentary use*, and its shift from being conceived as a device complementary to the landline phone to its becoming the absolute protagonist of the scene, *overtaking the landline.*

The mobile phone, and here we arrive at the fourth important variant, has also been the occasion for modern-day individuals to reconfigure public spaces by introducing a *new dynamic between the private and public dimensions*. The fifth variant, which concerns entrepreneurs, public administrations, and so on, is that these agencies have often found a way of making their subordinates use the mobile phone for work, *without paying for the costs*. In our list we now arrive at the most cited design variant created by mobile phone users (in particular, adolescents): the mass use of text messages, the use of *special ring tones*, again set for the purpose of communicating without paying, the practice of deliberately *calling random numbers* in order to widen one's social circle, the use of the mobile to *interact with the media* (for instance, voting during TV or radio programmes by means of text messages), phoning while in the car, and the use of the mobile phone *to present oneself.*

With its many uses, the mobile phone has become a fantastic resource for the construction of a public image, of self-assured behaviour, and to fill in gaps or empty moments with rituality. One variant reinforced by women is the importance attributed to the *aesthetic aspect* of the mobile phone. Moreover, a genuine design variant has been the attribution of *great importance to music* on the mobile phone. Starting from ring tones, but continuing with the possibility of listening to the radio on the mobile, young people in particular have been adopting this form of convergence between music and mobile phone. Finally, another design variant practised by users is the massive adoption of the pre-paid SIM (subscriber identity module) card.

The dynamics of the identity of the mobile phone have involved many societal changes: the changing perception of the private and public sphere (Haddon 1998; Höflich and Hartmann 2006; Katz and Aakhus 2002; Kim 2005), the increase of autonomy for adolescents and women (Ling 2004), the emergence of mobile phone-based mass mobilization and democratization processes in many countries (Nyíri 2003), the development of communicative dissonance and assonance, the relationship between cultural identities and mobile personalities (Nyíri 2005), the supporting and maintaining of communities and networks (Brown *et al.* 2001). Particularly interesting is the proposal to interpret the situational fluidity of the mobile sociability in terms of new 'floating worlds' (Gergen 2002, 2005) and to interpret it in terms of networked individualism (Wellman *et al.* 2006). Recently, Ling and Pedersen (2005) also studied how mobile phone users have transformed both the backstage and front-stage of social communication, involving changes in language, the psychological dimensions of mobile communication, and the presentation of self.

According to Sombart (1911), each culture domesticates and integrates a technology in its own way. Is this also the case for the mobile phone in China? In this country, is the mobile phone used to cut traditional schemas on sociability or to strengthen them? There are early works on this terrain trying to establish an agenda for future studies (see, for instance, Beaton and Wajcman 2004; Leung and Wei 1999; Miyata *et al.* 2005; Yu and Tng 2003).

The works of Law and Peng (2006) on migrant workers are most valuable for understanding how this segment of the Chinese population is using this device, for which purposes and with which meanings. From their works it emerges that migrant workers intensively use the mobile phone both to find new jobs and to keep alive family and kinship relationships, such as the bond between parent and child after separation following migration. The social consequences are that migrant workers are able to play a dynamic role in the labour market but also that traditional relationships otherwise threatened by the population mobility can survive and even be reinforced (B. Yang in this volume, Chapter 9).

Law and Peng also analysed the fundamental role played by the use of the mobile phone during the SARS (severe acute respiratory syndrome) epidemic (Law and Peng 2007: 137). It was in fact the use of the mobile phone among citizens getting and asking for information about SARS that obliged the Chinese government to admit that in China there was a serious health problem.

Supported and guided by this literature, we set up an explorative and qualitative research that aims to investigate the attitudes, opinions, expectations, and behaviours of users in respect to the identity of the mobile phone, and the social consequences of its use in China. In this chapter, after the brief review of the literature above, we will first discuss the aims and methods of the study. We will then focus on the results related to the identity of the mobile phone in China. This will be followed by a section analysing the sociability of the mobile phone, particularly 'semi-socialized individualism', and the high level of ritualization connected to its use. Finally, we will reflect on the overall results to draw a few tentative conclusions and pose open questions for future research.

Aim and methods

The research we present here is a qualitative study conducted with a convenient sample of 19 Chinese people. The sample consists of 8 males and 11 females, including college students, professors, white-collar workers, and migrant workers. Although all of the respondents lived in Beijing at the time of the interviews, they originally came from all over China, as can be seen in Table 10.1.

Our aims are essentially two: first, to explore how these users construct the identity of the mobile phone; and second, to investigate the paths and rituals of social change that have occurred in relation to the spread and adoption of this device. At the same time, we also try to analyse the users' attitudes, opinions, expectancies, and behaviours during these social processes.

The methodological instrument used is the semi-structured interview. The interviews were conducted in Chinese and then transcribed and translated into English. In so doing, we recorded respondents' basic socio-demographic data such as age, gender, education, income, professional position, and family organization. We also probed into how the identity of the mobile phone is constructed socially and how the mobile phone plays a role in supporting, maintaining, and/or opening new social relationships. Finally, we explored their practices in using the mobile phone, including episodes connected to the ritualized experience of our respondents with this device.

Results

The identity of the mobile phone in China

Three elements emerge from these interviews regarding the identity of the mobile phone in China.

1. *The mobile phone as a technology of writing*

The first element is that the mobile phone is especially used by our interviewees as an instrument of written communication, to send Short Message Service (SMS) text messages rather than to orally chat on the phone. So, for these respondents

Table 10.1 Profile of respondents

Name	*Profession*	*Gender*	*Age*	*Education*
Chunchua	Employee of an enterprise specializing in system integration	F	28	Major in Automatic Control Engineering
Fei (from Hebei)	Undergraduate student of Chinese Youth Political University	M	22	Major in the Department of Youth Education
H. Yan	Undergraduate student of Bei Hang University	M	22	Major in Software Studies
Hong (from Chongqing)	Editor of *Internet Weekend*	F	28	Bachelor degree from Peking University
Hongxia (from Xinyang-Henan)	Cleaner at property management company in Changwa District	F	41	Illiterate
Hui	Migrant worker working in a mine	M	22	Senior student from Liaoning Province
Hui	Editor of *China Development Brief*	F	24	Bachelor degree in Industrial and Commercial Business Administration
Jia (from Suzhou)	Undergraduate student of Chinese Youth Political University	F	22	Senior student at the Department of Public Administration
Jie (from Guangzhou)	Hakka broadcaster at China Radio International	M	22	Bachelor degree in Chinese from the Guangdong Meizhou Jiaying University
Li (from Hebei)	Works in a foreign-capital enterprise	F	34	Electrical engineer
Ping (from Wuxi)	Undergraduate student of Chinese Youth Political University	M	24	Major in Social Work
Qi (from Bei Hang)	Undergraduate student of Bei Hang University	M	25	Sophomore
Qiuying (from Hebei)	Works in the Real Estate Management of Changwa	F	26	Did not finish junior high school
Rong (from Beijing)	Professor at Tsinghua University	F	28	MA from the UK
Ruining	Official at the Party Committee of the Ministry of Agriculture	F	24	Bachelor degree
Ting	Undergraduate student of Chinese Youth Political University	F	22	Sophomore in the Department of Social Work and Management
Xiongwei (from Zhejiang)	Postgraduate student of Peking University	M	24	Major in Environmental Sciences
Yan (from Neimenggu)	Soong Qingling Foundation	F	24	Bachelor degree from Chinese Youth Political University
Yen (from Shanghai)	Undergraduate student of Chinese Youth Political University	F	22	Junior in the Law Department

the mobile phone seems to be, above all, a technology of writing, a technologically more developed pen. In general, the construction of the mobile phone as a technology of writing seems to be a design variant produced by users of all ages in China, whereas in Europe SMS was initially mostly used by adolescents. The reason for using the mobile as a technology of writing might be understood as an attempt to control and reduce communication cost. So such use of the mobile phone is very probably the only modality compatible with the current level of salary in China (Qiu 2009).

But there is another aspect that needs to be stressed: the fact that SMS allows more discretion, especially when privacy is involved (Song and S. Yang in this volume, Chapter 11). One of the consequences is that writing SMS messages can become a very tiring activity, given the amount of messages sent per day by several of our respondents. For example, Ruining sends between 20 and 30 SMS messages per day. Ping sends up to 500 SMS messages per month, even messaging for a few hours, which 'makes my fingers painful while inputting words'. 'Now,' confessed Chunchua, 'I'm very lazy and tired of typing on the cell phone, especially when the message is long.'

These statements make evident the reason why the use of the software *Paopao* (which facilitates the writing of words) to send SMS messages is very popular in China nowadays. From a semiotic point of view, it is interesting to analyse the paradox of mobile communication in China. The mobile phone is mainly used as a technology of writing in a situation in which the written language is very complex (Kadooka 2005). This implies that mobile writing has become very repertoirized in China. Here, in fact, there are many manuals for writing customized SMS messages for every occasion. Collections of jokes are also popular among mobile and Internet users (Lin in this volume, Chapter 12). So far, the high degree of formalization of SMS content seems to be peculiar to Chinese culture.

2. *The integration between the mobile phone and the Internet*

The second element that emerged is the integration between the mobile phone and the Internet. For our respondents the world of the mobile phone seems already to be quite integrated with the computer and the Internet, especially for the college students and white-collar workers whom we interviewed. Jie told us that he was saving his girlfriend's SMS messages on a website that allowed him to store information. His girlfriend did the same by building a website for his messages.

The computer is also useful when one needs information to make decisions when purchasing a mobile handset. Searching for information on technological products is one of the most widespread Internet practices. In other words, website operators have been among the first to create large communities of users, to whom they furnish a range of services and information.

But in China, unlike Europe, there is a more complementary use between the two media thanks to the wide diffusion of chat software for the mobile phone. In Europe, the new 3G service that supports such mobile chatting has been less successful (Rivière 2005). The converged mobile phone–computer in China can, however, provide such service via software like QQ, which is the Chinese version of ICQ. This chat software is usually available on the Internet, but one may also connect to it through mobile phone subscription. This habit of text chatting was shared by almost all the respondents. Xiongwei stated that he preferred to chat with his classmates, while Chunchua said, 'I used to chat with strangers for one or two years.'

The mobile phone seems to catalyse the need to enlarge sociability in many different ways: chatting with strangers, like in China, might be one; calling numbers at random, like in Italy, is another (Mazzara 2004). Mazzara, for instance, in his research on adolescents in Italy, discovered that this practice of calling numbers at random was very common among his respondents. The declared purpose was to make new relationships outside the inner circle of relationships that make up the mobile sociability (Leung 2001).

For our Beijing respondents, and their mirror respondents in European research projects, sending SMS messages or chatting seems to be far more popular than sending Multimedia Messaging Service (MMS) messages. Chunchua, for instance, said that she rarely sent MMS messages. Neither did Yan, because, she noted, they were still too costly (0.9 RMB), but she likes to take pictures of beautiful girls with her camera phone although she has a digital camera.

Clearly, camera phones require the Internet for the storage of the images produced. This tendency toward the integration of the mobile phone and the Internet is widely promoted as the newest service from the telecoms industry. Nevertheless, the storage capacity of mobile phones is improving, as Yan emphasized: 'Now my cell phone can save many messages, about 500. So, I saved many messages of humour and festival greetings and also all the messages from my boyfriend, which can be a record of our relationship.'

This bridge between the mobile phone and the Internet exists in a solid way among these respondents, though with different degrees of intensity. Despite this integration of mobile phone and computer, however, the mobile phone remains dominant. An individual's core affective relationships pass through the mobile phone, which makes the mobile the prominent medium. Above all, the mobile phone is used to chat.

3. *The mobile as fashion object*

The third finding is the relationship between mobile phone and fashion, in China (Yu and Tng 2003) as well as Europe. In both cases, women especially care about the aesthetic aspect of the technology. Indeed, aesthetics used to be an important part of technological objects in the pre-industrial age, as shown in the finely crafted telescopes, pistols, and watches of the time. An effect of mass production after industrialization was therefore to reduce these objects to their functional skeleton. In this sense, the connection between mobile phone and fashion can be seen as the return of the aesthetical aspect to technological objects, to please feminine customers in particular. For Chunchua, a female respondent, the design and also the brand of the mobile phone are quite important. 'I liked its small size [LG],' she added. It was very clear that female interviewees attached great importance to the design and brand of the mobile phone, particularly at the moment of purchase. Yan, above all, stressed the importance of the brand:

> Is appearance or function of the phone more important for me? I don't care about that. My second cell phone is a Chinese brand, the first colour-screen cell phone I used. I didn't trust Chinese brands from then on. The quality and service were really bad. So I think the brand is important, and also post-sale services . . . It would automatically power off. And the warranty was very bad. It had two colour screens, one on the front, the other inside. My classmate broke the front screen. Then we brought it to the shop for repair, which was in Zhongguancun Street. Someone there asked us to go to Dongsihuan [far away]. After we arrived there, they let us go back to the shop where we bought it. Then I was tired of getting it repaired.

Ruining told us that she changed her first mobile phone because 'it was too big'. The second one she chose looked 'like a tootsie' and was advertised as 'elegant, princely and beautiful'. When she bought this new mobile phone, she considered the design as the most important aspect. She said: '… the selling point to women is whether the mobile phone is beautiful. Other things are secondary. I always see whether it is beautiful first.'

Another respondent, Jie, took into account three things when he bought a mobile phone: price, battery capacity, and operational convenience. He said that 'many cell phones with colourful displays were beautiful but not good for use'. Ping, also a male respondent, commented on the fact that girls buy phone pockets, chains, and stickers: 'Girls treat their mobile phone as a pet.' Then he added, 'I do send a lot of SMS. But girls send SMS to the extent that they even treat the cell phone as one of the most important things in their life and hold it anytime and anywhere.'

Qi said: 'Some girls regard the mobile phone as a kind of decoration. There are many adornments for them, such as MP3.' Qi was also convinced that boys were more interested in the function while girls were more interested in the design of the mobile phone: 'It's like buying clothes,' he said. For him and other boys, the appearance and function of the mobile are two sides of the same coin, but they do not go to one extreme, unlike the girls, or so they thought.

In fact, several male respondents showed their appreciation for the design of the mobile phone. Xiongwei told us that, in the case of his new mobile phone, he was attracted by the appearance. 'It had a large screen, very comfortable to look at. And it's not expensive to me.' Meanwhile, many respondents complained about the ugly design of Nokia mobiles, while another respondent explained, 'If you go for design, Samsung, for instance, is the brand. But it is too expensive. There is no Samsung phone that's less than 2,000 yuan.' Another respondent, H. Yan, talking about his choice of a Mitsubishi mobile, said: 'I never thought of buying a domestic Chinese brand: the quality may not be good. I found this model: its screen was wide and had a fashionable appearance.'

These interviews make it clear that there is a tendency for male users to appreciate more the functional aspects of the mobile, while female users tend to appreciate the design aspect more than anything else. But, at the same time, fashion-related attributes of the mobile phone are becoming important for men as well, as has happened with clothes.

The sociability of the mobile phone

4. *The mobile phone and semi-socialized individualism in China*

In the Western context, scholars have talked about 'networked individualism' (Miyata *et al.* 2005; Wellman *et al.* 2006), in which the continuous development of individualism is supported by new technologies. Our approach focusing on the mobile phone might entail another metaphor: semi-socialized individualism. This metaphor seems to us to better express the sociability created by the use of this device in China. In China, individualization is a complex process, which has led to traumatic consequences due to rapid social and economic change brought about by globalization, industrialization, internal migration, the breaking-up of families, and new patterns of consumption.

Let us compare Canada with China. Wellman and Hogan (2006) found that East Yorkers in Toronto had nine very close members of their personal communities and 14 somewhat closer members in China. Generally when we investigated the social sphere around a mobile phone in Europe as well as in China, the number of members immediately increased. In this, as in other research projects, we often found respondents who claimed they had many tens of mobile telephone numbers stored in their mobile phone, of their friends, relatives, acquaintances, clients, and colleagues. In this research, Jie, for example, told us that he stored 200 mobile phone numbers in his mobile phone.

From this research it emerges that in China, more than in Europe, people would try to maintain relationships among schoolmates from primary school to university (Linshan and Thireau 2005). Constraints on sociality in the Chinese context are connected to the difficulty of creating and maintaining social networks in such a huge society. These difficulties are currently rapidly increasing because large-scale internal migration makes it more difficult to foster casual and spontaneous sociability around individuals, and obliges them to build their social network over and over again. In this situation all the patrimony of social relationships built during one's lifetime must be kept alive.

Schoolmates become part of the inner community of an individual. This is true not only in China but also in Korea, as Kim underlined (2002: 69). School-based social support and networking is so essential that in Hong Kong's print media 'mobile phones from the same brand are also referred to as *tong mon* [school fellows]' (Yung 2005: 362). Through a very tight social integration process, both mobile and Internet communication (though more especially mobile communication) play a central role in keeping alive and periodically revitalizing these relations, which, in addition to blood and regional ties, form the crucial network of relationships in China. It is a matter of fact that, without a mobile phone and/or computer, people living in dormitories would be beyond communicative reach and for this reason would fall into the black hole of non-existence. This was borne out by Ruining, who said: 'When I enter the office, the first thing is to check the mail, the second thing is to enter the college mates website, the third thing is to enter some websites for women.' And by Ping: 'I bought my new mobile phone because I need to contact my schoolmates.'

Through semi-socialized individualism, the Chinese respondents also reported increasing flexibility in making appointments, a phenomenon already well known in Europe. Yen said:

> With a cell phone you have an advantage: when you have an appointment with someone else, you don't have to determine the exact location or time in advance. You can look for your friend when you go there. But without a cell phone you have to decide everything in advance.

Another important phenomenon in China is the loss or theft of a mobile phone. These are very common events among our respondents and they accepted them with a kind of resignation. Chunchua told us that her mobile 'was stolen on a bus' and her reaction was that 'I was distressed'. Naturally, the loss of a mobile phone causes a great deal of trouble. Jie said that, if he forgot his mobile phone, he becomes nervous (cf. Vincent 2003). Ruining lost one mobile phone and another of hers was stolen near the gate of the school, which, together with Chunchua's story, indicates a level of crime in China that would have been unimaginable not too long ago.

5. *The mobile phone and a high level of ritualization*

Unlike in Western countries, where the issue of mobile phone etiquette is much studied, in China this aspect generally is harder to detect (Yung 2005: 352). But, among our respondents, the mobile phone created a new process of ritualization in the domestic setting of everyday life: for example, the first and the last gesture of the day might be to send a 'good morning' or a 'good night' SMS message. But there is another ritualistic practice related to the mobile phone, which is reported by Linshan and Thireau (2005) and is emblematic more than any other example. One of their respondents, a migrant worker, said that her parents had set up incense sticks behind the telephone and asked gods to protect her when she goes far away. Also between lovers who do not live together, the mobile phone has a strong capacity not only to penetrate the everyday rhythm of time, but to enhance new forms of control and to provide a psychological aid, thus allowing users to feel a sense of certainty. According to Ruining, 'I put it on the bedside when I go to sleep, and I'm woken up by the tinkling in the morning. I feel a bit steady and sure.'

This ritualization of time is shown also by the number of SMS messages people send to relatives and friends on the occasion of festivals. These include traditional ones such as the Spring Festival, the Dragon Boat Festival, the Festival of Lanterns, and Mid-Autumn Day. Other occasions may include celebrations of political connotation such as Women's Day, more commercially oriented festivals like Saint Valentine's Day and Singles' Day (11 November), or an inter-religious gala such as Christmas. Such ritualistic use of the mobile phone seems to be further popularized by a particular service, the volley, which allows one to send the same message to many different people at the same time. For this reason, the volley is probably more successful in China than Europe, where the service is generally used by young people for the purpose of group appointments. However, as individuals gain more freedom in their everyday life, some become intolerant of this kind of ritualization. As our respondent Ruining said, 'It is very boring.'

1. THE MOBILE PHONE AS TIMEPIECE

Inside this framework of ritualization, an interesting element that has emerged in China is the mobile phone's function as a measure of time. Jia uses the mobile phone as a watch and she is convinced that 'nowadays cell phones are more important than watches, which can be neglected'. Qi and H. Yan also use their mobile phone as a watch and alarm clock. H. Yan added: 'I don't wear a watch, because it is inconvenient and uncomfortable.' But of course these responses must be read cautiously, given the small sample size.

Nevertheless, they do seem to confirm another ethnographic research carried out in Italy, where it was found that, although watches are still common in the adult world at a mass level, they are no longer so popular among adolescents (Fortunati and Cianchi 2006). The watch is an old individualized technology that modernity has constructed as a necessity for citizens. But among the adolescents we interviewed (who are in higher secondary school), there were few wearing watches. Most interviewees spoke of their watches in terms of a generic present received for their confirmation, birthday, or Christmas. It seems that the watch is often perceived as costly and connected with the ritualization of traditional celebrations, which, however, does not correspond to the present trends in young people's everyday consumption behaviours.

In these cases, the watch seems to be considered by adolescents as a residue of their previous identity as children. It no longer serves to construct an image in the stage of adolescence. But this does not necessarily mean that watches are replaced by mobile phones. On the contrary, the spread of the mobile phone has implied a kind of redistribution of roles within the extended family of personal technologies. Youth in several cases prefer to use the mobile as a watch and alarm clock due to their limited financial means. When the financial situation improves, they may rediscover the watch and appreciate its ornamental aspect (Abdou 2004).

2. BATTERY

Another element of the social ritualization connected to mobile phone use is the battery. As Abdou underlines (2004), batteries provide the main source of electricity for mobile personal technologies. They are the doors that allow for mobility access. At the same time, Abdou maintains, they are archetypes of trivial objects whose presence is not put in question by individuals. The increase of mobility and the diffusion of mobile technologies, and especially of mobile phones, has inevitably led to increasing dependency on these power sources. Many forms of social ritualization in connection with worries about the short life of batteries emerged from the interviews. For example, users must recharge the battery of their mobile phone every evening, otherwise during the following day it does not work. This dependence on batteries, metaphorized as 'human energies' in the Hong Kong press (Yung 2005: 365), is one important source of disturbance for mobile users, like Jia, who said, 'The thin battery is almost a waste, you have to recharge it and soon no electricity is left.' Jia added that, in order to extend battery life, she had to turn off her mobile every night before sleeping. To avoid such problems, Ping and other respondents told us that they bought two batteries.

Conclusion

From this research we learnt that the identity of the mobile phone among our respondents has three main characteristics. First, the mobile phone is mainly used as a technology of writing, for psychological, social, and economic reasons. Second, the mobile phone is used more as a computer because, in the practices of using this technology, the convergence between the mobile phone and the computer/Internet might be more advanced than in Europe. Finally, the fashionable quality of mobile phone design strongly affects its purchase and use.

The third aspect refers to a process that has shown similar development both in Western countries and in China. The first and second aspects, however, describe a metabolization of this technological artefact, which seems to outline a Chinese specificity. The Chinese culture of the mobile phone seems to present itself as a third way between Europe, which has built in the first stage the great success of the mobile phone all over the world (Brown *et al.* 2001; Ling 2004; Lorente 2002), and Japan, which has developed the mobile phone in the most advanced and sophisticated way (Hjorth 2003; Ito *et al.* 2005; Harwit in this volume, Chapter 6).

Of course these findings are not that generalizable, but this exploratory study will serve as an initial point of departure, especially with this idea that China may constitute a third way of the diffusion, use, and co-construction of the identity of this device.

A second track for further research is to test the idea of semi-socialized individualism. The impressive diffusion of this technology has produced in China what we call a 'semi-socialized individualism', characterized by a great number of contacts in addition to a few closer or somewhat closer relationships and a reduced communicative modality.

The role of the mobile phone has to be understood in the context of profound social and economic changes characterizing China today. In many cases, the technological device seems to have supported the preservation and maintenance of traditional social structures. But will this trend continue in the future? There are in fact clear signals of new behaviours in the main social structures of China such as the diffusion of parallel love affairs and the violent creation of new paths of autonomy (Law and Peng 2007; Yung 2005).

Another important aspect is the highly ritualized use of the mobile phone with regard to time in the domestic setting as well as the ritualistic behaviours in taking care of the mobile phone, for instance in battery charging. Here we arrive at the third track for further investigation and exploration: the various forms of social ritualization. The mobile phone has not only inserted itself in the rhythm of existing rituality but also created new rituals of its own. Is the mobile phone used more to reinforce or to reorganize everyday life rituality? Further empirical studies and theoretical reflections would be in order.

Note

1 An early version of this study has been published as: Fortunati, L. and Yang, S. (2008) 'Identidad y Sociabilidad Móvil: el caso de China', in Juan Miguel Aguado and Inmaculada Martinez (eds) *Sociedad Móviles: tecnología, identidad y cultura*, Madrid: Biblioteca Nueva, pp. 137–165.

References

Abdou, J. (2004) 'Nouvelles technologies et énergie mobile. Ethnographie de l'usage des piles et des batteries', *Consommations et Sociétés*, 4. Online. Available: www.consommations-societes.net (accessed June 2006).

Baudrillard, J. (1968) *Le système des objets*, Paris: Gallimard.

—— (1972) *Pour une critique de l'économie politique du signe*, Paris: Gallimard.

Beaton, J. and Wajcman, J. (2004) *The Impact of the Mobile Telephone in Australia*. Online. Available: www.assa.edu.au/Publications/mobilephone.pdf (accessed June 2006).

Brown, B., Green, N. and Harper, R. (eds) (2001) *Wireless World: social and interactional aspects of the mobile age*, London: Springer.

Contarello, A., Fortunati, L., Gomez Fernandez, P., Mante-Meijer, E., Versinskaya, O. and Volovici, D. (2008) 'ICTs and the human body: an empirical study in five countries', in E. Loos, L. Haddon and E. Mante-Meijer (eds) *The Social Dynamics of Information and Communication Technology*, Aldershot, Hants: Ashgate.

Douglas, M. and Isherwood, B. (1979) *The World of Goods*, New York: Basic Books.

Eco, U. (1968) *La struttura assente*, Milan: Bompiani.semiologia. Turin: Einaudi.

Fortunati, L. and Cianchi, A. (2006) 'Fashion and technology in presentation of the self', in J. Höflich and M. Hartmann (eds) *An Ethnographic View on Mobile Communication*, Berlin: Frank & Timme.

Fortunati, L. and Manganelli, A.M. (2003) 'From the GSM to the UMTS: is it a path towards women?', in C. MacKeogh and P. Preston (eds) *Strategies of Inclusion: gender in the information society, vol. II: experiences from private and voluntary sector initiatives*, Trondheim: Center for Technology and Society, NTNU.

—— (2008) 'The social representations of telecommunications', *Personal and Ubiquitous Computing*, 12(6): 421–31.

Fortunati, L., Katz, J. and Riccini, R. (eds) (2003) *Mediating the Human Body: technology, communication and fashion*, Mahwah, NJ: Erlbaum.

Gergen, K.J. (2002) 'The challenge of absent presence', in J. Katz and M. Aakhus (eds) *Perpetual Contact: mobile communication, private talk, public performance*, Cambridge: Cambridge University Press.

—— (2005) 'Self and community in the new floating worlds', in K. Nyiri (ed.) *A Sense of Place: the global and the local in mobile communication*, Vienna: Passagen Verlag.

Gershenfeld, N. (1999) *When Things Start to Think*, New York: Henry Holt.

Glotz, P. Bertschi, S., and Locke, C. (eds) (2005) *Thumb Culture: the meaning of mobile phones for society*, Bielefeld: Transcript Verlag.

Haddon, L. (ed.) (1998) *Communications on the Move: the experience of mobile telephony in the 1990s*, Stockholm: Telia.

Haddon, L., Mante, E., Sapio, B., Kommonen, K-H., Fortunati, L. and Kant, A. (eds) (2005) *Everyday Innovators: researching the role of users in shaping ICTs*, Dordrecht: Springer.

Hjorth, L. (2003) 'Cute@keitai.com', in N. Gottlieb and M. McLelland (eds) *Japanese Cybercultures*, New York: Routledge.

Höflich, J. and Hartmann, M. (eds) (2006) *An Ethnographic View on Mobile Communication*, Berlin: Frank & Timme.

Ito, M., Daisuke, O. and Matsuda, M. (eds) (2005) *Personal, Portable, Pedestrian: mobile phones in Japanese life*, Cambridge, MA: MIT Press.

James, W. (1890) *The Principles of Psychology*, 2 vols, New York: Henry Holt.

Kadooka, K-I. (2005) 'A contrastive study of messaging and orthography in English, Chinese and Japanese', paper presented at the conference Seeing, Understanding, Learning in the Mobile Age, Budapest, April.

Katz, J. (ed.) (2003) *Machines that Become Us*, New Brunswick, NJ: Transaction.

Katz, J. and Aakhus, M. (eds) (2002) *Perpetual Contact: mobile communication, private talk, public performance*, Cambridge: Cambridge University Press.

Kim, D.S. (2002) 'Korea: personal meanings', in J. Katz and M. Aakhus (eds) *Perpetual Contact: mobile communication, private talk, public performance*, Cambridge: Cambridge University Press.

—— (2005) *When Mobile Came: the cultural and social impact of mobile communication*, Seoul: Communication Books.

Latour, B. (1998) 'Fatti, artefatti, fatticci', in M. Nacci, *Oggetti d'uso quotidiano. Rivoluzioni tecnologiche nella vita di oggi*, Venice: Marsilio.

Latour, B. and Woolgar, S. (1979) *Laboratory Life*, London: Sage.

Law, P.L. and Peng, Y. (2006) 'The use of mobile phones among migrant workers in southern China', in P.L. Law, L. Fortunati and S. Yang (eds) *New Technologies in Global Societies*, Hackensack, NJ: World Scientific.

—— (2007) 'Cell phones and the social lives of migrant workers in southern China', in R. Pertierra (ed.) *The Social Construction and Usage of Communication Technologies: Asian and European experiences*, Diliman, Quezon City: The University of Philippines Press.

Leung, L. (2001) 'College student motives for chatting on ICQ', *New Media & Society*, 3(4): 483–500.

Leung, L. and Wei, R. (1999) 'Who are the mobile phone have-nots?', *New Media & Society*, 1(2): 209–26.

Ling, R. (2004) *The Mobile Connection: the cell phone's impact on society*, San Francisco, CA: Morgan Kaufmann.

Ling, R. and Pedersen, E. (eds) (2005) *Mobile Communication and the Re-negotiation of the Social Sphere*, London: Springer.

Linshan, H. and Thireau, I. (2005) 'Liens personnels, expressions, repères d'identification. Actions expressives et nouveaux supports de communication en Chine', *Réseaux*, 133: 69–99.

Lorente, S. (ed.) (2002) *Juventud y teléfonos Móviles*, special issue of the journal: *Revista de estudios de Juventud*, n.57.

Maldonado, T. (1997) *Critica della ragione informatica*, Milan: Feltrinelli.

Mazzara, M.B. (2004) 'I giovani e il telefono cellulare: nuovi contesti e nuovi stili di interazione sociale', paper presented at the national conference of the Italian Sociological Association, 'Lo studio del mutamento sociale: problemi metodologici e ricerche empiriche', Salerno, January.

Miyata, K., Wellman, B. and Boase, J. (2005) 'The wired – and wireless – Japanese: webphones, PCs and social networks', in R. Ling and P. Pedersen (eds) *Mobile Communications: re-negotiation of the social sphere*, London: Springer.

Nyíri, K. (ed.) (2003) *Mobile Democracy: essays on society, self and politics*, Vienna: Passagen Verlag.

—— (ed.) (2005) *A Sense of Place: the global and the local in mobile communication*, Vienna: Passagen Verlag.

Qiu, J. (2009) *Working-class Network Society: communication technology and the information have-less in urban China*, Cambridge, MA: MIT Press.

Rivière, C. (2005) 'Mobile camera phones: a new form of "being together" in daily interpersonal communication', in R. Ling and P. Pedersen (eds) *Mobile Communications: re-negotiation of the social sphere*, London: Springer.

Silverstone, R., Hirsch, E. and Morley, D. (1992) 'Information and communication technologies and the moral economy of the household', in R. Silverstone and E. Hirsch (eds) *Consuming Technologies*, London: Routledge.

Sombart, W. (1911) 'Technik und Kultur', *Archiv für Sozialwissenschaft und Sozialpolitik*, 33: 305–47.

Vincent, J. (2003) 'Emotions and mobile phones', in K. Nyíri (ed.) *Mobile Democracy: essays on society, self and politics*, Vienna: Passagen Verlag.

Wellman, B. and Hogan, B. with K. Berg, J. Boase, J. Carrasco, R. Cote, J. Kayahara, T.L.M. Kennedy and P. Tran (2006) 'Connected lives: the project', in P. Purcell (ed.) *The Networked Neighbourhood*, London: Springer.

Yu, L. and Tng, Hou T. (2003) 'Culture and design for mobile phones for China', in J. Katz (ed.) *Machines that Become Us*, New Brunswick, NJ: Transaction.

Yung, V. (2005) 'The construction of symbolic values of the mobile phone in the Hong Kong Chinese print media', in R. Ling and P.E. Pedersen (eds) *Mobile Communications: re-negotiation of the social sphere*, London: Springer.

Part IV

Exploring the Chineseness of mobile phone use

11 College students' self-positioning and mobile phone consumption[1]

Jing Song and Shanhua Yang

Introduction

In recent years, mobile phones have become increasingly popular amongst college students in China. Based on a series of interviews with college students in Beijing, this chapter examines the way in which college students position themselves through their consumption of mobile phones; we refer to this as 'self-positioning'. Self-positioning involves an understanding of the self (or self-identification) on one hand, and the evaluation of the socio-ecological context on the other. Based on this analysis we gain an understanding of the consumption patterns of these college students.

As far as self-identification is concerned, most students have to rely on their family for financial support, which is associated with their constrained consumption and careful rationing of mobile phones. Meanwhile, college students often have relatively high self-expectation, and consuming luxury goods matters less for their current social status compared with other social groups. College students also use mobile phones in unique ways to define their interpersonal relationships by identifying the in-group and the out-group while reserving their personal arenas.

As for the socio-ecological context, explicit or implicit regulations on campus contribute to some latent rules that constrain the use of mobile phones by college students. But campus life also provides access to other communication technologies that can supplement or substitute that of the mobile phone. Moreover, the use of mobile phones among college students is subject to peer influence, and becomes part of the campus culture. In the final analysis, an attempt to describe mobile phone usage and its social impact on college students in China should take cognizance of the way in which a particular group of college students position themselves within their particular social contexts, and this should be set against campus life in China as a whole.

Previous studies have shed light on changes in the self resulting from the use of mobile phones, for example based on a new conception of space and time, but they also notice that diverse personal factors may shape the way the mobile phone is used (Garcia-Montes *et al.* 2006). This study tries to add to the investigation of the dialectical relationship between the self, the social environment, and the mobile phone by focusing on the consumption behaviours of college students.

Issue: the penetration of the mobile phone in campuses

The mobile phone has been penetrating the China consumer market at great speed in recent years. Within that market, as an important consumer group, college students have attracted increasing commercial attention. In this investigation, we interviewed ten college students: two came from Peking University, five from China Youth University of Political Sciences, and three from Beijing University of Aeronautics and Astronautics. Five of them were undergraduates at the time of the study, and the other five were graduate students.

The interviewed graduate students entered undergraduate college around 1997 and 1998, when public phones were the only means of long-distance communication on campus besides letters, and they witnessed great changes in the dominant information and communication technology. These interviewees recalled that users often queued up to use public phones on campus, until each dormitory room began to have a line phone during 1998 and 1999. Since 2000, the colleges and universities in Beijing started to establish the campus LANs (local area networks), which facilitated the use of the Internet by students. According to respondents, mobile phones began to emerge among students around 2000 as well. At the time of the interviews, mobile phones have been prevalent among both graduates and undergraduates. A female sophomore reported that more than half of the students in her class own a mobile phone, whereas a male junior stated that 90 per cent of his classmates have one. In a class contact list at China Youth University of Political Sciences, only two students left blank the item 'mobile phone number'. A similar tendency was discovered based on a survey of Taiwanese college students, when the mobile phone evolved 'from a luxury for businesspeople into an important facilitator of many users' social relationships' (Wei and Lo 2006: 53).

The proportion of mobile phone holders appears to be higher among graduate students than among undergraduates. But beyond the age effect, the timing of the first purchase of a mobile phone among graduate students and undergraduates reflects a cohort effect. Most interviewed undergraduates bought their first mobile phones during their second year in college, whereas the interviewed graduates purchased their mobile phones much later in general. Among five graduate respondents, two purchased mobile phones in the junior and the senior year, and the other three purchased mobile phones after they entered graduate school (one in the first year and the other two in the second year of graduate school). Undergraduate respondents started using mobile phones at an earlier stage, as well as their cohort. An undergraduate respondent said that some of her classmates bought mobile phones in their first year in college, most of the rest began to use mobile phones in the second or third year, and at the time of the study almost all the students in class had a mobile phone. A graduate student commented that 'the undergraduates nowadays are so different from the way we were'.

This 'purchase lag' between cohorts of students lends support to a marked increase in mobile phone consumption during 2002–2003 among the population of university students, partly due to large-scale reductions in price in the mobile phone market. A respondent recalled that the price of a Nokia 3310 mobile phone fell from more than 3,000 yuan in 1999 to 1,000 yuan in 2002, and to 800 yuan at the time of the study (see also Wang and Cheng in this volume, Chapter 5). In addition to the drop in the prices of mobile phones, there were also various SIM (subscriber identity module) card sales promotions, such as the 'Travel in China' card, the 'M-zone' card, or that of 'Global Communication', accompanied by deals in service packages (see discussion below in the sub-section 'Identification on economic status'). As a result, more students are attracted into the mobile phone market.

Mobile phone consumption and self-identification

As the use of mobile phones and the sending of SMS (short message service) messages has become part of the daily experience of many college students, their lives may have undergone an individualization process, as has been witnessed in many other consumer groups. However, the mobile phone consumption pattern amongst college students is largely embedded in their shared characteristics and experience, and also reflects the continuous process of self-identification. Self-identification involves their recognition, construction, and expression of identity in different areas, including economic status, self-expectation, and interpersonal relationships.

Identification on economic status

The students' attitudes to consumption are influenced in the first place by their assessment of their own economic situations, and that of their families. College students do not have regular income, there has been a continuous increase in tuitions that used to be very low before the 1990s, and the market-oriented reforms also came along with the rise in living costs. According to respondents, the tuition and stipend add up to 3,000–10,000 yuan per year. Accommodation is one of the major expenditures, and most college students live in dormitories. For instance, at the China Youth University of Political Sciences, students live either in cheap dormitories or in more spacious apartments at a higher rent. The graduate students in Beijing University of Aeronautics and Astronautics pay 2,000 yuan per year for their dormitories and facilities. Graduate students are in general more independent economically from their families, because they receive a small amount of subsidy on a monthly basis from the government, and they are more likely to be engaged in part-time work or paid research activities. An interviewed graduate student at Beijing University of Aeronautics and Astronautics earned 400 yuan monthly as a teaching assistant, but his family still provided for 60 per cent of his daily spending.

As such, the parents' attitudes count a lot. Six interviewees reported that their parents had encouraged them to buy mobile phones in order to facilitate communication between family members. Furthermore, all except one respondent drew on the family purse to buy or to use mobile phones, and the economic dependence is more pronounced among young and female students. Two female undergraduates got new mobile phones from parents as a gift. Another female undergraduate received a used phone from her father, and similar practices of 'shared' usage are said to be common among her friends, one of whom gave her mobile phone to her mother in exchange for her mother's new-design mobile phone. The age and gender differences and how the students got their mobile phones may suggest different kinds of economic bonds within families, which are also illustrated by who pays for the use of mobile phones. For instance, a father gave a mobile phone to his daughter together with a pre-paid SIM card so that he could pay for his daughter's mobile phone bills. These practices may relieve students of their economic obligations but these students feel that their mobile phone consumption becomes a burden to the family. As a result, half of the interviewees reported that they would keep using the same mobile phone and postpone the purchase of a new one until they could afford to buy it themselves after finding a job.

In short, because of their limited budgets and an awareness of their economic situations, college students have to consider how to economize in their consumption of mobile phones. For our interviewees, affordability was a prior concern in the decision-making process. Though new models, with features such as colour screens and chord ringtones, were attractive and considered to be 'cool', their higher prices proved prohibitive to some interviewees. The mobile phones used by our interviewees are neither particularly plain and cheap, nor especially elaborate and flashy, with the prices ranging from 899 to 2,000 yuan, and mostly between 1,000 and 1,500 yuan. Two of the students reported that they had obtained their mobile phones at exceptionally low prices (899 and 900 yuan), and their strategy was to patiently pay attention to sales promotion information from the Internet.

Gender difference also stands out in the unique strategies students employ to find cheap mobile phones. The information-searching behaviour was frequently reported by our male interviewees, who tended to make decisions after collecting enough information from various sources such as the media and comparing available model designs in different shops, which was further facilitated by their access to the campus Internet system. Three male interviewees suggested to us that the younet.com website was informative on the mobile phone market, which helped them to find information on promotions or discounts. In contrast, female students tended to employ the information-exchange strategy and choose mobile phones based on comments and recommendations by parents, classmates, and friends.

College students not only search for cheap mobile phones but also conduct careful trade-offs between cost and quality of service packages related to SIM cards to make the best choice. Three of the ten respondents have persisted in their use of 'Travel in China', another three have been using 'M-zone', and the rest have changed between these two back and forth. Both choices are attractive to some extent: the 'M-zone' packages charge a fixed monthly fee for service with an upper limit (e.g. 20 yuan for 300 SMS messages maximum, or 30 yuan for 500 messages), while the 'Travel in China' charge is based on the actually rendered service at a higher rate without monthly fee. Competing with each other, providers often offer additional deals for their service as well. In the promotional activities, the rate of long-distance voice calls on the 'Travel in China' card was reduced from 0.6 to 0.49 yuan per minute by dialling a '12593' service number before the destined number, while that of the 'M-zone' card dropped to a price range of 0.2 to 0.4 yuan among 'M-zone' users, and 'M-zone' consumers can make up to 35 minutes of free voice calls every month. All in all, special promotional offers surely became one factor taken into consideration when students made their choices.

By taking advantage of various service packages, college students manage to spend less on mobile phones. The 'M-zone' service packages attracted many college students during 2002 and 2003 because of the good deals of sending and receiving messages as allowed in the service, and thus became a good choice for those interviewees who switched from more costly voice calls to message-chatting. In comparison, the service provided by the 'Travel in China' card seems to be more economical for those who do not send text messages frequently. This is the reason why one interviewee changed back to the 'Travel in China' card after a trial period with the 'M-zone' service. In short, various SIM cards succeed in enticing different consumer groups because their unique services match different consumption needs effectively. Instead of following advertisements that claim to offer the most economical service packages, college students deliberately find out the option that fits their understanding of their own needs.

In spite of the fluidity in consumption choices, college students also have concerns about the sunk cost and they would not change from one SIM card to another too frequently. At the same time, college students continue to pay attention to the ways and means for reducing mobile phone costs, such as using line phones as a supplementary way of communication in case of long-distance calls, and substituting voice calls for text-sending, thus explaining why SMS usage is of such great volume in China. In our data, most interviewees reported their monthly costs on the use of the mobile phone, including voice calls and SMS messages, to be between 40 and 60 yuan; two students spent more than 100 yuan per month; and one other managed to limit monthly expenditure on the mobile phone to 10 yuan.

Identification on self-expectation

Colleges and universities are places of prestige in contemporary China. College students anticipated a higher chance of upward mobility, as university education is in general regarded as related with higher socioeconomic status in the future. This perception is associated with college students' self-image, and consequently has an impact on their mobile phone consumption. On one hand, college students may feel that there is no necessity to prove their socioeconomic status while they are still at a tertiary-level educational institution, regardless of the amount they have available for the purchase of mobile phones. On the other hand, their choices of consumption can be influenced by anticipated consumption needs after graduation.

Because of expectations about their future requirements, college students are, as one of the interviewees pointed out, unaffected by the need to 'show-off' (see

Peng and Chu in this volume, Chapter 13). Interviewees reported that their purchase of a mobile phone is based on 'when there was a need', and none of the interviewees regarded ownership of a mobile phone as a means of enhancing self-esteem. Although not owning a mobile phone would be inconvenient, our interviewees indicated that it would not embarrass them. Furthermore, the mobile phones that our interviewees used were cheaper in general compared with those found among migrant workers in Guangdong. In the project conducted in Guangdong, it was found that a sense of actively controlling technology brings about a certain psychological satisfaction, which balances the fact of the low economic status and minimal living environment of migrant workers and helps to rebuild their self-esteem (Chu and Yang 2006).

These findings suggest that the self-fulfilment needs of college students are not achieved through the symbolic significance of consumption goods; these needs are met by other means.

The relatively high expectations that university students have for themselves result in a balancing between their anticipated future consumption needs and their current consumption needs. In addition, the development of mobile phone technologies and applications make it expensive to constantly stay abreast of new products. Furthermore, new models and functions are sometimes considered unreliable, and some interviewees expect that people will need to pay less for them in the future. They find it an exhausting exercise trying to keep up-to-date with fashion and technological advances, and a conservative strategy is to delay consumption. Three students said they would buy a better mobile phone after graduation, and their plans on future consumption added to their awareness of the necessity for curbing their present desire in favour of consumption in the long run.

In addition to the desire to curb current expenditure in favour of future plans, the mobile phone consumption among college students is also affected by their consumption plans on other products such as MP3 players, digital cameras, and PCs. Many of these digital products have similar functions with the advanced models of mobile phones, but some interviewees said that they preferred specialized digital products with more reliable qualities, instead of mobile phones that have add-on functions to store data, to take pictures and videos, to play music, to connect to the Internet, and so on. They reported that they would postpone consumption and buy specialized products later rather than buying an all-in-one mobile phone now. Being confident that they will be able to afford more expensive specialized digital products, they believe that the delay in enjoying such functions would be repaid. Some male interviewees drew on their knowledge on information and communication technology (ICT) and explained to the researchers that all-in-one designs have the downside that the quality in providing each single aspect of service is likely to be undermined. For example, they may encounter problems when transferring data from mobile phones to PCs because of format incompatibility, and so on. These concerns contribute to their conscious assessment of consumption choices.

Deliberate consumption plans among college students can be seen as a reaction to similar trends characterized by the ever-expanding array of gadgets observed in Western societies, which has given rise to the fear of being dehumanized by personal communication technologies. On one hand, these technologies 'become' extensions and representatives of people, and on the other, these technologies 'become' integrated with people's clothing and bodies (Katz 2003). College students seem to have a particular view of how technologies can be part of their lives and what it can attach to their self-worth. Based on their educated judgements and the awareness of potential problems, college students tend to be sober and careful consumers.

Identification on interpersonal relationships

Though the values college students attach to the use of mobile phones are mainly instrumental, mobile phones play a vital role as the extension of the self. With the help of mobile phones, college students find themselves better equipped in expanding social networks, pursuing market opportunities, promoting career development, and dealing with emotional issues. More than that, mobile phone communication can restructure one's social circles and redefine barriers of personal arenas. Similarly, Wei and Lo find that, among Taiwanese college students, the mobile phone 'supplements the fixed telephone as a means of strengthening users' family bonds, expanding their psychological neighborhoods, and facilitating symbolic proximity to the people they call' (Wei and Lo 2006: 53). In other words, mobile phone communication is a way where college students work on their social ties and in-groups based on their understanding of who they are or want to be.

Our interviewees use mobile phones to broaden or narrow the social circles that they identified with by carefully selecting how and with whom to communicate. The contacts they saved on their mobile phones are mainly classmates and friends in addition to family members. Whenever someone on the contact list calls them, they are able to recognize who is calling immediately and thus can decide to answer the call or not. In addition, college students use mobile phones to differentiate between 'strong ties', such as close friends and family, and 'weak ties', such as acquaintances (Granovetter 1973), based on the frequency of communication. In particular, partners are among the most contacted ones according to the respondents who have one.

The communication methods also make a difference. According to some interviewees, compared with direct voice calls, SMS messages are considered a casual, cheaper, and 'safer' way of communication with someone that you do not know very well. One can send the same message to many receivers at the same time, which can be effective in maintaining one's weak ties. The interviewees also received messages that were sent to many people, like weather forecasts and sportscasts. A large proportion of the 'copied' messages that are repetitively circulated are greeting messages, particularly on occasions of festivals and birthdays. Such 'copied' messages can sometimes be supportive in periods of loneliness and depression. When the 2003 Severe Acute Respiratory Syndrome (SARS) epidemic broke out, college students were advised to stay within campus to avoid being infected, and those who were suspected of carrying the SARS virus were required to be diagnosed to decide whether they should be hospitalized. Some interviewees, intimidated by the danger underlying physical contact with other people, turned to the mobile phone and cyber communication to contact others. It helped to establish a 'distant presence' for those isolated under protection measures, even if the greeting messages and those of 'health tips' or 'prescriptions' are 'copied' messages. However, sometimes our interviewees felt that they were flooded by 'copied' messages and they chose to prioritize originally composed messages, because 'genuine affection cannot be copied'. For example, one interviewee said that he did not appreciate the 'copied' jokes sent from their friends as a substitute for a greeting.

Originally composed messages play an important role in our interviewees' lives, strengthening their strong ties. College students in general are good at inputting Chinese characters using mobile phone keyboards and more creative in generating combinations of symbols to convey subtle information, and many of them are experts and big consumers of messages. Two students choosing the 20 yuan user package said it was not enough even though they could send 300 messages per month. Another student added a 5 yuan service to the former user package, which enhanced the number of messages he could send by 120. Those who chose the 40 yuan user package were accustomed to sending more than 500 messages a month. Moreover, the sending and receiving of messages is a silent act, which fosters a sense of privacy able to allay the concerns of users (Yang and Chu 2006). A female interviewee said that, compared with texting, walking out of the dormitory room and making a voice call is another way to secure privacy, but it can be offensive to her roommates as a signal of mistrust. Under circumstances such as the interviewee who had quarrelled with her roommates and wanted to complain to an outsider, texting is also preferred as an 'unvoiced' means of communication. In addition to privacy concerns, texting also helped in romantic relationships. Because most college students do not have a private room, texting is a good channel for emotional expression; two interviewees bought mobile phones in order to contact their girlfriends more often, particularly using messages. In the selective sending and responding process, SMS messages play an important role in helping college students to get closer to or keep a distance from someone.

Gender difference is also observed in choosing how and with whom to communicate using mobile phones. Female interviewees are more aware of the potential danger of being harassed when receiving a call from an unfamiliar number. One female student said that she would delete the missed-call records from strangers instead of calling back. Male students, however, tended to be more carefree in answering incoming calls. However, mobile phones can be used for protective purposes as well despite the risks they bring about. One female interviewee used it to slow down the pace of getting closer to a cyber-chat mate on OICQ (the name of one cyber-chat software application meaning 'oh I seek you'). The first step was to exchange mobile phone numbers, and the interviewee said that she would not have done so if the cyber-chat had made her uncomfortable. The second step was to exchange SMS messages for a while until she felt it was time to exchange line phone numbers. After they talked on the phone, they eventually met.

The interviewee considered it unwise 'to tell someone who you just met on-line the number of your dormitory phone', by which personal information may be leaked, and the potential harassment may upset other roommates as well. Texting using mobile phones provides a safer option. In other words, mobile phones can be used to strengthen relationships and to keep a reasonable distance when a virtual encounter is turned into a relationship in the real world.

Similar ways of mobile phone use are found in daily interaction as well. Some interviewees chose to exchange mobile phone numbers instead of line phone numbers to avoid disclosing the physical location of their dormitory. Another advantage of leaving mobile phone numbers is, when they are contacted, mobile phones can be carried to places of solitude in order to protect the privacy of conversations. But mobile phones bring other risks. For example, others may peek at saved calls and messages. To protect privacy, two interviewees even destroyed their old SIM cards when changing to a new service system. Furthermore, two of the interviewees discovered that a default setting allows other mobile phone users under the same service system to track their locations, and they had to request the company to stop this service immediately. In other words, mobile phones can be empowering but also deprive people of control over information flows. Though mobile phone users can choose when and with whom they communicate, they can always be contacted now, and some interviewees complained about the 'always-on-line' pressure. Three interviewees said they hated to be 'reachable' at any time, but turning off the mobile phone is also offensive and may cause them to miss important calls. Some interviewees suggested that one should delay the purchase of mobile phones until it was absolutely necessary, and the contact list should be expanded with caution. The convenience of using mobile phones lets one be capable of relating to more people, but also challenges the way people view themselves and their relationships (Gergen 1992), which made our interviewees think twice about their self-identity.

Socio-ecological context and mobile phone consumption

Together with the perception that college students have of themselves, the socio-ecological context surrounding them is another factor closely related to their consumption choices. As instruments to manage their campus life, the use of mobile phones is restrained by explicit and implicit rules on campus, and influenced by the needs of college students to communicate with their parents, classmates, professors, potential employers, friends, and partners.

The campus is a unique public sphere where faculty and students need to locate themselves in flows of time and space segments scheduled for coursework and other activities. Even in the dormitories, students share space with their roommates, and there is limited 'private' time and space for each student. With relatively rigid noise constraints, students are expected to turn their mobile phones to silent mode in class, and many of them change the ring type to the vibrate mode in classrooms. One male interviewee also did so in the dormitory to avoid disturbing others. In many cases they are not supposed to answer incoming calls but to wait until later to call back. Our interviewees found that 'texting', as a silent act, was less constrained by time and place, though a female interviewee still insisted it was better to reply to messages after class except for urgent situations.

Interviewees who use mobile phones in a polite way expect others to be respectful as well. They felt outraged when they kept being disturbed by beeps and rings generated by mobile phones in libraries and classes, especially during times like the reading period before the final exams. Some interviewees were also aware that they could be distracted by their own mobile phones, so one interviewee decided to keep his mobile phone off when he was preparing for a test. In sum, compliance with the campus learning environment affects the duration, location, frequency, and method of mobile phone usage.

Another factor of the socio-ecological context that affects the consumption of mobile phones is the availability of campus facilities that can be alternatives to mobile phones. As discussed above, mobile phones appeared later than the establishment of the dorm line phone system (1998 in Beijing University of Aeronautics and Astronautics and 1999 in Peking University), and even before the dorm phone was available, students had access to phones in offices, laboratories, or other public places on campus. Though the rise of mobile phones has created an alternative to the line phones, some interviewees pointed out that line phones have certain advantages. Interviewees from Peking University reported that both the two line phone systems in most dormitories managed to introduce cheaper services. The rate for long-distance calls on the '201' system was reduced from 0.6 to 0.4 yuan per minute if the user dialled '17909' first, and the 'tietong' system, which was put into use in 2002, also charged less compared with the conventional line phone system. One interviewee used the line phone to make a long-distance call to his parents every week and it only costs him around 30 yuan per month.

College students have also been exposed to the Internet longer than to mobile phones, not only because of the dorm LAN system (since 2000) but also due to the fact that libraries, offices, research centres, and laboratories on campus are mostly connected on-line. Some graduate students who spent most of their time in laboratories are able to use lab computers all day long and they do not need to pay for cables, hubs, and high-speed servers. According to a graduate student from Beijing University of Aeronautics and Astronautics, the laboratory where he worked paid a monthly 200 yuan for each Internet Protocol (IP) address to guarantee that everyone in the lab can be well 'connected'.

The establishment of the LAN system, however, allows students to use a PC almost everywhere on campus (particularly in the dormitory) to access the Internet with a minimal monthly fee (10 yuan for students at CYU and 5 yuan in PU). Besides receiving and sending emails, college students can remain 'on-line' using message manager tools such as OICQ and MSN as well as chat rooms. Many college students are also frequent users of alumni websites such as Chinaren that aim to help people find their alumni and old friends, as well as the internal Bulletin Board System (BBS) that acts as a bulletin board for students on campus. One interviewee, a graduate student from BUAA, regarded BBS to be a more convenient way of communication than email.

To sum up, line phones and the Internet are substitutes for mobile phones to a great extent. In using mobile phones, both parties are charged for voice calls, and despite its adjustment in the few years before the time of the study, it still costs more to use mobile phones than line phones in general. To cut the cost, two male interviewees report that their parents only call their mobile phones if they cannot be reached by dorm phones. One student regretted that he answered all the incoming calls during the first month of using a mobile phone, which cost him 400 yuan, almost equivalent to his monthly non-accommodation stipend. After that he chose to respond to incoming calls from mobile phones by using the line phone or text messages. Some interviewees use the line phone specifically to communicate over a long distance and mobile phones for contacting friends who are in the same city.

As such, if one uses the mobile phone wisely, particularly in combination with the line phone and other ways of communication, considerable cost can be saved. One interviewee used to call his former classmates in other cities using the 201 line phone system, which cost him thousands of yuan a year. Turning to the combination of SMS messages and line phones helped him to cut the cost of distant calls substantially. Another female interviewee developed a more complicated strategy to combine the advantages of mobile phones and line phones to save costs. To make the most out of different communication devices, this interviewee reached a 'consensus' with her parents: she will not pick up calls from her parents on the mobile phone, but either call back and hang up soon to indicate that she is in the dorm and can be reached by the dorm phone number, or give no response when she is out, and call back and hang up again when she is 'ready' after going back to the dorm. When the interviewee wants to contact her parents, she simply calls her father and hangs up before her father answers, and soon her parents will call her using the dorm phone. In such ways, she kept her mobile phone cost as low as 200 yuan a semester.

Meanwhile, mobile phones cannot be replaced in many aspects in college students' lives. In a study of young people's use of mobile phones in America, interviewees 'cleverly use the landline numbers for certain business transactions and keep the mobile phone numbers to those who are their in-group members' (Aoki and Downes 2003: 358). But in our study, some interviewees chose to provide mobile phone numbers in both cases, such as when they were about to enter the job market and submit job applications. Because the private space is limited on campus, and college students only stay in the dormitory or an office equipped with a line phone within certain periods, they prefer to be contacted by mobile phone as it allows more privacy and flexibility. Mobile phones are also used to register for on-line games and services and to schedule appointments and reminders. But as one interviewee points out, the most popular use is to receive and send short messages because 'one can write to anyone at any time in any place'. The sharing of daily experience through text messaging added to the popularity among college students, which allows some researchers to view mobile phone use in a fresh way, noting that 'young people use mobile phone content and the phones themselves to participate in the practices of gift exchange' (Taylor and Harper 2003: 267).

But even the most popular aspect of mobile phones – texting – has some disadvantages compared with e-mails, Internet chat rooms, and voice calls. Some interviewees suggested that SMS messages were not always reliable and sometimes too short and casual for formal settings. Because of the chance of delayed and mislaid messages, three interviewees pointed out that a voice call worked better on either serious or urgent matters, such as communicating with their supervisors. One interviewee said that, to explain something complicated, it often took her four or five messages. Although the maximum length of a message keeps being extended and some mobile phones are able to store more than 100 messages, SMS communication is still not comparable with e-mail with its attachment facilities.

Furthermore, though SMS messages are easy to pick up by college students, their parents or other older people may have difficulty in mastering inputting symbols, letters, and Chinese characters using the small mobile phone keyboard. For those who are used to the PC keyboard and can access the Internet most of the time, Internet communication is more comfortable and cheaper, though it only costs 0.1 to 0.15 yuan for each SMS message. Therefore one interviewee actually considered it unnecessary to have a mobile phone when LAN was available for use on campus. However, the popularity of PCs and mobile phones has resulted in an emerging connection between the two kinds of communication among college students. Some interviewees told us that some Internet chat rooms, such as *NetEase Bubble*, provide a service to link users' on-line account with their mobile phone numbers, so that people can talk to their friends on-line using PCs, and off-line using mobile phones. Some on-line game websites also send confirmation codes to their users' mobile phones to confirm the users' ownership of the virtual identity before the service is provided so that anonymous users cannot 'pretend' to be others, harass other users, and invade cyber databases. These new developments in technology are based on the fact that more and more college students are both PC users and mobile phone consumers.

Last but not least, mobile phone consumption is influenced by information flows and popular attitudes on campus, particularly those from the peer groups that students identify with. In an increasingly diversified market, young people often use consumption to redefine themselves, which helps to shape an emerging brand-conscious youth culture in China (Wang 2005). The choice of mobile phone is such an example: though often affected by accidental factors, there are several brands and models that are relatively popular among college students. Among our interviewees, four of them have been using Nokia and another four have stayed with Motorola, and the rest of the interviewees have tried brands like Samsung, Alcatel, Sony-Ericsson, Mitsubishi, and Eastcom. A female interviewee from Beijing University of Aeronautics and Astronautics said that, in her dormitory, three out of four roommates bought a Nokia mobile phone one after another. In addition to the choice of mobile phone, that of SIM card is also influenced by peers. One may choose to follow the trend when most of his or her contacts change to one type of SIM card (and the related service package), not only to enjoy the 'insiders' benefits provided by the service package but also to relocate oneself in a community. One female interviewee did so when many of her friends changed to the 'M-zone' card. According to another female student, she not only uses a similar service system as her friends but also set up specific ringtones for incoming calls from close friends.

It is noteworthy that most interviewees stated that their choice is based on independent deliberation and some interviewees were not comfortable being regarded as 'following' fashion. Though some studies find a strong influence of trend-setters exerted over girls' consumer behaviour in their class or group, under which 'the right clothes, the right mobile phone and the right leisure accessories are her ticket of admission to the group and confer status' (Tufte 2003: 77), our interviewees as college students, both male and female, seem to value the identity of being a deliberate and independent consumer, though they are often receptive to their friends' comments at the same time.

To conclude, the mobile phone consumption of college students is related with how they evaluate their socio-ecological context and how they position themselves in the social milieu of the campus. In this way, college students are able to make the most out of the highly commercialized and diversified ICT equipments. Based on the self-positioning processes in real and virtual environments, college students gain a better understanding of their advantageous and vulnerable aspects as a customer in the mobile phone market.

Conclusions

College students are an important component of mobile phone consumers. Their consumption patterns are greatly affected by the way in which they position themselves. This process of self-positioning involves how they perceive themselves as individuals and how they evaluate themselves as individuals in a social context, more especially in the context of campus life which has an increasing affinity with cyberspace.

There are four aspects of college student identification which have a bearing on mobile phone consumption:

1 College students identify themselves as non-income earners, and their consumption choices are largely based on information about prices.
2 College students have high expectations that their future careers will be self-fulfilling. These expectations affect their current consumption patterns, which are not influenced by a need to prove their self-worth through the possession of a mobile phone and, in particular, the possession of an up-to-date model.
3 College students appear to be selective in defining their social circles, and their concern to preserve personal arenas also affects their mobile phone consumption choices.
4 Self-positioning also involves an understanding of the socio-ecological context in which students position themselves. That context needs to take into consideration the demands that campus life makes on the students. First, the use of mobile phones must be tempered by campus timetables and the demands of the learning environments. Second, the Internet, dormitory phones, and

other public phones serve as alternatives to mobile phones. E-mail and the Internet are formerly regarded as communication instruments, and access to a combination of various communication devices is popular amongst students. Third, the choice of models and service is influenced by peers and this in turn has an effect on interpersonal relations and social circles among students.

As mobile phones permeate campus life at an unprecedented rate, college students have begun to accept them as tools for managing their personal information and their social lives. The use of mobile phones provides them with an opportunity to present their identities, to redefine consumption needs, to expand social networks, and to preserve personal arenas. Yet mobile phones also expose college students to different kinds of tension in the transition to adulthood that are related to economic constraints, information overload, and the abstraction of the real world into virtual interactions. College students are in the process of learning how to handle these challenges which cannot be easily solved by relying on the development and diffusion of technology. The strategies to cope with the mobile-phone-related daily experience employed by college students arise from how they position themselves in their growth trajectory, in the campus environment, and in a globalizing modern world. Being affected by self-identification and environmental factors, the mobile phone consumption pattern of college students can be used as an example of how people deal with technology-facilitated interactions in an age of information and of uncertainty.

Note

1 This chapter is a modified version of a research paper based on a research project conducted in March 2004, in which ten students were interviewed from Peking University, China Youth University of Political Sciences, and Beijing University of Aeronautics and Astronautics. The research collaborators include Liu Xiaojing from CASS, as well as Peng Yinni, Yang Ke, and Jiang Qin, who were then research students at the Sociology Department of Peking University.

References

Aoki, K. and Downes, E. (2003) ‘An analysis of young people’s use of and attitudes toward mobile phones’, *Telematics and Informatics*, 20: 349–64.

Chu, W.C.R. and Yang, S. (2006) ‘Mobile phones and new migrant workers in a South China village: an initial analysis of the interplay between the “social” and the “technological”’, in P.L. Law, L. Fortunati and S. Yang (eds) *New Technologies in Global Societies*, Hackensack, NJ: World Scientific.

Garcia-Montes, J., Caballero-Munoz, D. and Perez-Alvarez, M. (2006) ‘Changes in the self resulting from the use of mobile phones’, *Media, Culture & Society*, 28(1): 67–82.

Gergen, K. (1992) *The Saturated Self: dilemmas of identity in contemporary life*, New York: Basic Books.

Granovetter, M. (1973) ‘The strength of weak ties’, *American Journal of Sociology*, 78(6): 1360–80.

Katz, J. (2003) *Machines that Become Us*, New Brunswick, NJ: Transaction Publishers.

Taylor, A. and Harper, R. (2003) 'The gift of the gab? A design oriented sociology of young people's use of mobiles', *Computer Supported Cooperative Work*, 12(3): 267–96.

Tufte, B. (2003) 'Girls in the new media landscape', *Nordicom Review*, 1: 71–78.

Wang, J. (2005) 'Youth culture, music, and mobile phone branding in China', *Global Media and Communication*, 1(2): 185–201.

Wei, R. and Lo, V. (2006) 'Staying connected while on the move: mobile phone use and social connectedness', *New Media and Society*, 8(1): 53–72.

Yang, S. and Chu, W.C. (2006) 'Shouji: quanqiuhua beijing xia de "beidong" jieshou he "zhudong" xuanze – zhusanjiao diqu nongmingong shouji xiaofei de wenhua he xintai de jiedu' [Mobile phone: passive acceptance and self-initiated choice in the process of globalization], *Guangdong shehui hexue yinjiu* [Social Sciences in Guangdong], 2: 168–73.

12 Romance and sexual ideologies in SMS manuals circulating among migrant workers in Southern China

Angel Mei-yi Lin

Introduction

Law and Peng (2004, 2006, 2007) have reported on the widespread use of the mobile phone and Short Message Service (SMS) among migrant workers in Southern China. While there were only 85 million mobile phone users in Mainland China in the year 2000, by mid-2004, the number had reached 305 million. The national average penetration rate was 24 per cent, with the Guangdong province having an above-national average penetration rate of around 60 per cent. Migrant workers are among the lowest-income groups in Guangdong, taking up largely skilled and semi-skilled jobs in factories with monthly salaries ranging from 400 to 800 yuan (US$1 is approximately 8.23 yuan) (Law and Peng 2006). However, a migrant worker is willing to spend double or triple their monthly salary to purchase a mobile phone. The trend of increasingly cheaper subscription rates has turned these migrant workers into more frequent users instead of frugal users (ibid.). Law and Peng indeed describe a number of both social and practical functions served by mobile phone and SMS use among these migrant workers, including maintaining contact with family members in their home village far away from their workplace, maintaining contact and friendship with kinsmen/women (from the same village or kinship network) who are scattered in the Southern provinces, to maintain contact with ex-workmates, to get job-market information, and for purposes of dating and courtship (Law and Peng 2004, 2006, 2007).

Among the above-described purposes for using mobile communication, one important topic for research on gender relations among these migrant workers is the use of SMS among migrant workers for dating and courtship, for seeking romance and sex partners, and for experiencing romantic and sexual adventures and relationships. According to the ethnographic observations and interviews of Law and Peng, the migrant workers in Southern China readily make use of different kinds of SMS messages and have turned to SMS manuals in making their messages:

> Messages involving greetings, salutations, and humorous works are also popular among the workers. Most of these are written in rhymes. Some of these are modified from messages the workers have received, and some are

> from SMS message manuals that collect a great number of interesting SMS rhymes categorized into different topics. The manuals also serve to teach the users how to create emoticons and animations, compose ring tones, and use digits representing Chinese sentences. When workers receive interesting SMS rhymes or interesting animations, they will forward them to their friends so that they can enjoy the messages together in cyberspace.
>
> (Law and Peng 2007: 137)

With their relatively low literacy level from their limited schooling experience, these migrant workers seem to readily turn to the SMS manuals, which are published as handy, smaller-than-pocket-size booklets selling cheaply (e.g. 3 yuan) and which circulate widely among these migrant workers. These manuals are organized as lists of numbered entries of short message texts placed under different topic headings (e.g. 'Humorous Sex Jokes', 'True Sentiment Confessions', 'Witty Love Letters' and the like). The language styles of these entries vary a great deal even within the same manual and can range from colloquial, explicit, slangy Chinese rhymes to subtle, poetic and metaphoric literary language. The manuals seem to offer a textual supermarket of diverse linguistic and discursive resources made available to the migrant workers when they use SMS for purposes which are related to seeking (virtual) relationships and (virtual) experience of romantic/sexual pleasure and excitement.

Against this backdrop I present a textual analysis of a sample of these SMS manuals. In the following sections I shall first report the findings of the textual analysis and then discuss the implications of these findings for understanding romance/sexual ideologies and practices and gender relations among these migrant workers in Southern China.

A textual analysis of SMS manuals

In this study I analysed 19 SMS manuals purchased in the factory areas which migrant workers frequent in Dongguan, in Guangdong province. The titles analysed are listed in Table 12.1 (for ease of reading, only the English translations are included). The titles belong to the category of smaller-than-pocket-size manuals. In these publications no information on the editor can be found. These seem to be the most widely circulating type of manual among the migrant workers as they are handy, easy to carry and cheap (the cost is 3 yuan each). Some of the content in these manuals overlaps – some entries in one manual can be found in another manual.

The bulk of the messages in the manuals consist of two main types. The first type can be called sexual jokes and they seem to be more or less explicit erotic jokes shared by the message-sender/writer with the message-recipient/reader to poke fun or arouse some vicarious form of sexual pleasure or excitement through the graphic images conjured up in the message. The second type can be called romantic expressions of love and they are messages expressing the message-writer/sender's love for the message-reader/recipient. There are other sub-types

Table 12.1 SMS manuals analysed

1	Expert of Fun Making
2	Poems and Songs
3	Love of Heaven and Earth
4	Group of Fun Makers
5	Short Messages with Long-lasting Love
6	Mini Lover's Talk
7	Gentleman's Friendship
8	Humorous Jokes
9	Creative Jokes
10	Love Confession on the Seventh Day of the Seventh Lunar Month
11	Love
12	Men and Women in the Material World
13	Language of On-line Community
14	Clever Talk
15	Funny Lies
16	Chivalrous Frame of Mind and Tender Feelings of Love
17	Popular Password
18	Love Humour
19	Jokes among Internet Users

which can be seen as variations of the above two main types. Below I shall analyse each type with some illustrative examples. In the brackets after the citation of each example I shall indicate the manual title number (see Table 12.1 for full titles) and page number from which the example is found.

Sexual jokes

As mentioned above, this type of message seems to be meant for poking fun or for sharing a laugh between the SMS sender and recipient. Most of the jokes/

messages seem to be written from a male perspective and will be readily appreciated, recognized and enjoyed by a male reader. The subject positions projected in the messages seem to be predominantly male heterosexual. For instance, in many manuals, there is the topic heading of 'Erotic Humour' or similar headings such as 'Foolish Talk'. Some jokes/messages under this type of heading are written in the linguistic form of a rhyme, such as the one below found in Manual 5 under the topic heading of 'Erotic Humour' (for ease of reading, only the English translations are presented):

> Tongue twister of McDonald's: Two Big Macs in the chest, with two black grapes in front. There is a cleavage between two thighs; everyone roars in laughter after eating it.
>
> (Manual 5: 41)

This message plays on appropriating McDonald's food rhymes and using images of food items to suggest female body parts. Although one has to be familiar with the Chinese language to appreciate the rhyming phrases, the literal translation in English can give the reader a sense of the kind of sexual explicitness conveyed in graphic images as portrayed in the rhyming lines in the message. Playing on Chinese–English bilingual word puns is a characteristic of another sub-type of this kind of message. For instance, under the topic heading of 'Wordmonger' in Manual 10, we can find the following example:

> A pretty lady wearing a T-shirt with the number 99 on the chest; a foreigner wants to praise her, but he forgets how to say 'nine' in Chinese, so the foreigner says: Miss, your two 'nine-nine' are very beautiful!
>
> (Manual 10: 51)

The number '9' pronounced in English twice as in '9–9' sounds like the pronunciation of the word for 'tits' in Mandarin Chinese. The sexual image of the message will be conjured up as the reader activates the sounds of the numbers '9–9' as guided by the textual context of the message.

Some other erotic humour messages do not take the form of a rhyme or word pun but the form of a short story (narrative) with a punch line where the reader's sex-related expectations are contradicted or shattered:

> Last night I was desperately seeking you. In my naked body I longed for you; I couldn't stay in bed without you. Only you can make me feel good. Where are you? My sleeping gown!
>
> (Manual 5: 41)

Still another sub-type of this kind of message makes use of the narrative formula of beginning the message with expressions like 'I love you, I cannot live without you …' and then ending the message with a punch line which goes more or less like this: 'If you believe this you've been taken in' or 'I hope you won't believe this; I'm cheating you'. Below is an example from Manual 9 under the topic heading of 'Witty Love Letters':

> Thinking of you, in every minute and second; loving you, till the seas dry up and the rocks melt away; together with you, to the four corners of the earth; waiting for you, till there are giant earthquakes and landslides; hoping you … don't take this as real.
>
> (Manual 9: 27)

Although the above message is placed under the heading of 'Witty Love Letters' in the SMS manual, it hardly constitutes a sincere love letter and it seems unlikely that a male sender will send it to a female recipient without risking a breakdown of their relationship. It is thus more likely that this kind of pseudo love letter is used as a joke and shared perhaps mostly among male senders and recipients for fun. It might be the case that a male sender sends it to a male recipient posing as a female sending what appears to be a love letter to him, but then ends the message with the punch line which reveals that it is actually a joke. Another possibility is that these jokes are sent and shared among close male friends who will not be offended but will enjoy a good laugh from such a play on romantic love letters. Still another possibility is that these jokes are mostly consumed by male readers who can derive pleasure from merely reading them (i.e. not necessarily sending these jokes by SMS). Further ethnographic work on the actual uses of the different kinds of messages in SMS manuals is needed to shed light on the hypotheses made here based on the textual analysis.

Romantic expressions of love

In contrast to the above type of sexual jokes, within the same manual are messages expressing romantic notions of love and longing for the other. For instance, under the heading of 'Classical Charming Remarks' in Manual 4, there is a glorifying of love by associating love with angels, God and light:

> Love is the union of two persons' souls which have become an angel flying towards God. This angel will bring God the glory of the world.
>
> (Manual 4: 58)

Besides the romantic glorification of love which associates love with purity, light and sacredness, many messages under this category also express strong romantic, fateful notions of love. For instance, the message 'You are the love destined for me':

> As Heaven has destined us to meet in the great sea of people, in this life of mine I can no longer live without you.
>
> (Manual 16: 79)

The romantic notion of love destined by fate or made in Heaven, and realized in meeting the one unique lover that one cannot escape (i.e. one love, one life), is a recurrent theme in this type of message. For instance, below is an example under the heading of 'True Sentiment Confessions':

> By destiny, I met you; by fate in life, I liked you; driven by passion, I have fallen in love with you; I cannot help missing you. I will love you with all my heart. I love you!
>
> (Manual 8: 7)

This kind of message seems to encompass both the male and female perspective. It seems that this kind of message can be sent by both male and female senders and the subject positions projected by the message can be either male or female. Some romantic messages focus specifically on on-line love relations:

> Clearly knowing that being lovesick is bitter, stubbornly I am unable to forget you. Falling in love with you is not my fault; it is a disaster caused by the Internet.
>
> (Manual 1: 13)

> Behind the unreal Internet there are sincere, beating hearts. I know we cannot have a future, but I still want you to know: I love you.
>
> (Manual 12: 56)

Both of the above messages seem to indicate a kind of forbidden love or love without a future, but both indicate the strong romantic desire of the message writer for the other. It draws on the recurrent themes found in the rhetoric of romantic, forbidden love. The messages sound sincere and it seems that messages like these can provide good discursive resources for men who want to court on the Internet women who are married or who already have boyfriends (hence, forbidden love or love without a future). However, the messages can be seen as speaking from either a male or female subject position and can be used by both men and women to develop relationships on the Internet.

It can be seen that the romantic type of messages in these SMS manuals stands in sharp contrast to the sexual jokes. While these two types of message are the mainstay of SMS manuals, there are also a small number of messages which seem to express a diverse range of ideologies about romance, sex, marriage and modern-day gender roles and relations.

Minor types of messages

These messages are scattered under different topic headings which do not actually relate closely to the content of the messages. It seems that the editors of the manuals loosely assemble the messages under different headings. In the following sections, however, I have attempted to classify these messages under different headings, which are the researcher's categories and not the original headings in the manuals.

Expression of sexual desire and passion

This type of message differs from the sexual jokes described earlier in that it seems to try to express sexual desire and passion without making a joke of it. It seems to recognize and convey a serious sense of the legitimate naturalness of human sexual desire and passion:

> Love has both the spiritual and physical aspects, which should be equally important; otherwise, love is not complete, because we are not gods, and we are not beasts, either.
>
> (Manual 4: 72)

The above message presents an argument for the need to recognize both the spiritual and physical aspects of love (i.e. to have both love and sex as necessary components for love to be 'complete') based on the grounds that as human beings we are neither purely spiritual beings nor purely physical beings. The message acquires a rational, logical tone with its parallel argumentative structures: the message begins with the first statement that both the spiritual and physical aspects of love are important, supported by a second parallel two-part statement asserting that humans are neither gods nor beasts. The tone of the message strongly suggests an image of a rational male speaking subject. Sometimes there is a message which speaks in a very traditional female tone:

> Brother is a dragon in heaven, and Sister is a patch of flowers on earth. If the dragon doesn't turn its body, it won't rain; if there are no showers of rain on the flowers, the flowers won't blossom in red.
>
> (Manual 6: 56)

The above message strongly suggests a traditional Chinese female speaking tone with its use of traditional address terms for lovers and romantic couples: Brother and Sister. The message draws on a traditional saying in the last sentence: if there are no showers of rain on the flowers, the flowers will not blossom in red. The original idiomatic saying goes like this: if there are no showers of rain on flowers, the flowers will not blossom in red; if a person does not work hard, one will not succeed. However, by appropriating just the first part of the traditional saying as its concluding part, the message draws on the image of flowers turning red upon receiving showers of rain. The message also draws on the images of a dragon, rain showers, flowers and flowers blossoming in red to remotely signify the pro-active male role (to give) and the reception-oriented female role (to receive) in the sexual act. The use of 'Brother' and 'Sister' to signify lovers/partners also draws on traditional literary tropes in both serious and folk literature. The image of 'heaven' and 'dragon' as male, and 'earth' and 'flowers' as female, also draws on traditional Chinese gender hierarchies/ideologies. While the message suggests a female speaking subject who boldly, albeit metaphorically, speaks about the complementary male and female roles in the act of sex, it is also a reinforcement of a traditional Chinese woman's subjectivity on the issues of gender relations/hierarchies. Although the message seems to project a female speaking position, it is possible that the message can also be used as a sexually pleasurable piece of reading for male readers of these SMS manuals. It seems that a man can derive pleasure from reading messages like this which speak in a traditional submissive female's voice expressing a sexual desire for him.

Metaphors of love and marriage

Another minor group of messages expresses witty metaphors about marriage and marital relations in the format of wise sayings. Below are some examples which are diverse in their metaphors and yet quite explicit in their ideologies on love and marriage; for instance, the following message shows its clear ideologies about love and marriage, and men and women:

> Marriage is a keyboard, there are too many regulations and rules; love is a mouse which works with a little click; men are like the central processing units, the inside is most important; women are like monitors, showing you everything.
>
> (Manual 15: 77)

In the above message computer metaphors are used to contrast marriage with love. Marriage is compared to a keyboard that has too many rules and regulations whereas love is as quick and easy as a click. Stereotypical gender images are also presented: males are like the central processing units and appearance is not as important as their contents. Women are compared to computer monitors: everything is on and for display; a woman's appearance is more important than her inner qualities. The message reinforces the stereotypical images of women and men and of love and marriage. The following message further describes how women become tied down after marriage:

> A woman before engagement is like a swallow who flies in whatever way she likes; after engagement, she is like a dove who can fly but dare not fly too far; after marriage, she is like a duck who wants to fly but can fly no more.
>
> (Manual 12: 63)

All of the above messages convey negative views of marriage and stereotypical gender images. In the following section we shall look at examples of a small number of messages which focus specifically on images of gender and gender relations in modern-day societies.

Images of gender and gender relations

There are a small number of messages which convey ideological images of women, as the following message shows:

> Four ideals of women: money falling from the sky; all other pretty girls on earth have died; the brains of handsome guys got sucked; and all [the men] are crying and yelling to love them.
>
> (Manual 8: 106)

Women are portrayed as money-greedy, eager to spend money without having to make the effort to earn it, and wishing that there was no competition from other pretty women for handsome men. Such messages present an image of women that is shallow, materialistic and self-centred. Occasionally one might come across a message that presents an alternative view of women:

> Pride and self-confidence are a woman's best accessories. A woman without confidence and hope, even if she is not ugly, still does not possess the kind of attraction that moves people's hearts.
>
> (Manual 17: 58)

The above message presents an image of a self-confident woman and describes her as having the kind of attraction that will move people's hearts. It focuses more on the woman's inner qualities, rather than her outer appearance, as being more important and more attractive. There are some witty messages presenting some stereotypical images of men as well as gender and sexual relations, as the following examples show:

> Betrayal is men's blood, universal love is men's manifesto, freedom is men's byword, and looking for change is men's fashion.
>
> (Manual 6: 54)

> A magazine in Japan did a survey on men's activities after having sex, and the results are as follows: 5% smoke; 8% sleep; 87% get dressed and go back home.
>
> (Manual 6: 82)

The above messages convey ideas commonly circulating in folk discourses about men and women's differing expectations regarding love and sex relations: men focus on sex and women focus on physical appearance; men focus on changing partners, searching for new excitement and searching for virgins (or women who are inexperienced in love and sex), and cannot be loyal to one woman, whereas a woman focuses on becoming her lover's last woman, hoping that he will be loyal.

While the above examples show a range of ideologies, the mainstay of the SMS manuals is still the first two types of message described earlier: sexual jokes and romantic messages. In the next section we shall discuss the implications of the findings of the textual analysis.

Romance, sex and gender relations among migrant workers in Southern China

As mentioned above, the mainstay of these SMS manuals consists of sexual jokes and romantic messages. While the sexual jokes can be rather explicit and feature colloquial language, the romantic messages seem to aim at conjuring up beautiful romantic images and metaphors of love. The puzzle arises for the analyst to understand how these two very different types of message with extremely different ideological orientations can co-exist side by side in the same manual.

From the above textual analysis of the messages we can see that most of these two types of message (especially the sexual jokes) seem to speak from a male subject position and perspective and address a male audience. It is then possible that most readers and consumers of these manuals are male migrant workers who use these manuals for several major purposes. The erotic images in the sexual jokes constitute messages that most men can readily recognize, decode and appreciate, and that they can derive pleasure from merely reading these erotic jokes. Some might like to share these jokes with their male friends through SMS; some might just read them for pleasure. This explains why the sexual jokes take up a lot of space in these SMS manuals, which also serve as pleasurable reading material for men. However, the question remains why the second type of message (i.e. romantic messages) also takes up a lot of space in the SMS manuals if the consumers of these manuals are mostly males who are focused mainly on gaining sex-related pleasure.

Another possibility is that while the male consumers of these manuals consume the jokes as sex-related pleasurable reading, they also use the manuals as resources for composing and assembling romantic messages to send through SMS to prospective female targets. Since women generally like to hear sweet words and receive romantic, comforting messages, the male consumers of these manuals might be drawing on the romantic messages in the manuals to compose their SMS messages.

In a study on the mobile communication of migrant workers in Southern China, Law and Peng write that, when a male migrant worker intends to court a female worker, he sends SMS messages to express his feelings towards the girl or indicate his desire to date her, and that nowadays, with the widespread use of mobile phones, young migrant workers more readily express their emotions, especially through SMS (Law and Peng 2006: 251). On the other hand, female migrant workers also seem to enjoy indulging in the romance offered by the virtual/SMS world. For instance, in the same ethnographic study, Law and Peng write about a female migrant worker who enjoys chatting on the Internet and through SMS, although she knows that the romance offered there is just an illusion:

> Another female worker told us that she enjoyed logging on to the chat rooms and chatting with male netizens although she had a steady boy friend. The chat rooms she logged on to were mainly designed for workers, for example 'Romance in the Workplace'. She said that she used different names, such as 'Apple', 'Ice' and 'Windie', when she logged on to different chat rooms. Usually, she would chat with the netizens for a while and if she found them interesting, they would exchange mobile phone numbers and keep in contact by sending messages. Her record for sending messages was 160 in one day to at least 7 different netizens. She knew that the sweet words and honeyed phrases were an illusion, yet she enjoyed chatting as she found their words comforting. Although she had found that relationships developed through chat rooms were not reliable, she had once dated a netizen working in the same village. She had, however, found it hard to develop their relationship because the mystical feeling achieved in cyberspace had disappeared when they met face-to-face. According to our information, this worker was having some problems with her boyfriend partly because he was not that romantic when they were together. It seems that she was looking for the romance on the network that she lacked in real life. Perhaps their problems will intensify if she continues meeting people on-line.
>
> (Law and Peng 2006: 254)

It seems that women migrant workers who find their real-life boyfriends not romantic enough readily escape into the Internet and SMS world to find the kind of romance they desire. The SMS manuals are full of these sweet messages for male migrant workers to draw on and use in this virtual world to develop romantic relationships with female migrant workers. However, as Law and Peng indicate above, these romances can just stay in the virtual world: they often shatter once the girl meets the boy in real life. SMS manuals in this sense serve a purpose for both male and female workers: they are like the romantic serials on TV (Ang 1985, 1996) or the romantic fiction (Belsey 1994) that women indulge in, but this time they have on-line male partners who play along with them in co-constructing their romantic virtual/SMS world.

Bourdieu (1998/2001) points out that, in contemporary European and American societies, men and women have very different points of view on love relations: men are inclined to compartmentalize sexuality. They often conceive it in terms of conquest, and also the sexual act itself is seen by men as a form of domination and possession of women. In contrast, women are often socially prepared to see sexuality as an intimate and emotionally highly charged experience which does not necessarily involve penetration but a wide range of activities (e.g. talking, touching, embracing, caressing). Bourdieu's observations about the differing expectations of European and American males and females in love and sexual relations seem to be also relevant to our study here. It seems that the male migrant workers are also more inclined to differentiate sex and romance and can derive great pleasure from reading sexual jokes, whereas female migrant workers seem to be more like the European/American females: they need romance and sweet words for their pleasure.

It also seems that the notions or ideologies of love and marriage circulating in these SMS manuals tend to be largely portraying the negative aspects of marriage (see the examples above). Romance for romance's sake seems to be a popular ideology. It seems that romance leading to marriage is not a preferred choice of either men or women (e.g. men do not want to be tied down by marriage; women know that men will not stay loyal or romantic in marriage). The virtual/SMS world provides an escape and a fantasy land for women who desire romance but cannot find it in real life. On the other hand, it seems that male migrant workers still readily send romantic messages to potential targets in their attempts to court them as they know that their female targets like such romantic messages.

This cat-and-mouse chasing game seems to be a paradoxical one as male migrant workers who tend to focus on sex-related pleasures, on changing partners and on finding new excitement in new partners will forever be frustrating the female migrant workers in real-life, face-to-face relationships. Female migrant workers, who are cognizant of the different ways of men (e.g. the tendency not to be loyal to only one partner, the tendency to become unromantic in real-life contact), will, however, still choose to indulge themselves by escaping into the Internet and SMS world which is full of romantic messages copied from or modelled on the romantic messages in the SMS manuals.

Perhaps the above interpretation can solve the analyst's puzzle raised at the beginning of this section. However, a long list of questions remains: Who are the consumers of these manuals and what do they use these manuals for? What are the contexts of consumption of these manuals? How and when do the migrant workers use them? Do different kinds of users use the different kinds of messages differently for different purposes? All these are important questions which cannot be answered by textual analysis alone. Further ethnographic work on SMS manual consumption practices is needed to answer the above set of questions. Despite this, the present textual study of SMS manuals contributes to the generation of research questions and hypotheses for informing and planning future ethnographic research into the role of mobile communication and SMS in constituting the often paradoxical relationships between men and women in rapidly industrializing and urbanizing societies, where traditional images, norms and practices of gender, gender roles and gender relations are undergoing processes of rapid-fire change and the institutions of marriage and family are also being transformed by rapid social and economic processes.

References

Ang, I. (1985) *Watching Dallas: soap opera and the melodramatic imagination*, London: Methuen.

—— (1996) *Living Room Wars: rethinking media audiences for a postmodern world*, London: Routledge.

Belsey, C. (1994) *Desire: love stories in Western culture*, Oxford: Blackwell Publishers.

Bourdieu, P. (1998/2001) *Masculine Domination*, Stanford, CA: Stanford University Press.

Law, P.L. and Peng, Y. (2004) 'Cell phone, Internet and the SARS epidemic: a case study in China', paper presented at the International Workshop on Mobile Technologies and Health: Positive and Negative Aspects, Udine, Italy, June.

—— (2006) 'The use of cell phones amongst migrant workers in Southern China', in P.L. Law, L. Fortunati and S. Yang (eds) *New Technologies in Global Societies*, Hackensack, NJ: World Scientific.

—— (2007) 'Cell phones and the social lives of migrant workers in Southern China', in R. Pertierra (ed.) *The Social Construction and Usage of Communication Technologies: Asian and European experiences*, Diliman, Quezon City: The University of the Philippines Press.

13 Mobile phone usage in Chinese society[1]

Yinni Peng and Rodney Wai-chi Chu

Mobile phones and society

Technological innovation has brought human society into a new era – the age of information. It is an age in which information and communication technologies (ICTs) have gradually become indispensable to modern life, and the adoption of them has initiated dramatic social change. As a result, more and more scholastic attention has been paid to the social changes that emerge from the use of ICTs. Sociological and communicational discussions concentrate mainly on the study of mobile phones and research surrounding the effects associated with the use of the Internet. In mobile phone studies, researchers (Kopomaa 2000; Fortunati 2002; Katz and Aakhus 2002; Puro 2002; Ling 2004; Katz 2008) believe that the use of the mobile phone has fundamentally changed the way people communicate with each other, and the way in which they organize their social lives. The concepts of time and space have also changed since mobile phones have penetrated into people's lives. Conceptions about the division of the public/private sphere have been blurred. Therefore, the use of the mobile phone has not only dramatically changed 'how people live their lives', but also 'how they see their world' (Katz and Aakhus 2002: 1).

However, in recent years, more and more researchers have studied the issue from the opposite direction: how the social context in which the technological innovation occurs influences its use (Haddon 2003; Kavoori and Chadha 2006; Leonardi *et al.* 2006; Mesch and Talmud 2008; Goggin 2008). As Haddon (2003: 47) said, 'how one experiences ICTs is not completely predetermined by technological functionality or public representations; it is also structured by the social context into which it is received'. According to the above scholars, culture is also regarded as an important factor which influences people's mobile phone usage and their interpretations of mobile phone usage. Thus, the usage of the mobile phone has increasingly been considered as an active adoption by our modern society rather than a passive reception. In this chapter, we aim to investigate how Chinese culture and social context modulate the use of the mobile phone in Mainland China.

China's mobile communication era was launched in the late 1980s when mobile phones were still regarded as luxury goods. After the late 1990s the number of mobile phone users began to increase substantially (Wang and Cheng in this volume, Chapter 5). According to statistics provided by the Ministry of Industry and Information Technology of the PPC between 1999 and 2004, a period of five years, the number of mobile phone subscribers multiplied eightfold from 43.3 million to 335 million (MIIT 2004). By December 2010, the number of mobile phone users in Mainland China had reached 859 million, a figure which represented a population penetration rate of 64.4 per cent (MIIT 2011). China has become the biggest national mobile phone market in the world. In the light of this cataclysmic expansion of the mobile phone market in Mainland China, it is of great importance to examine how the use of the mobile phone is interpreted by Chinese society.

This chapter examines how Chinese society itself shaped the use of the mobile phone. It points out that there are two aspects of Chinese interpretation which govern the distinctive features of mobile phone usage: (1) the situation-specific/relationship-specific usage of the mobile phone in the public sphere; (2) face (*mianzi*), which refers to image, social standing or impression created by the possession of items of apparel such as clothing or a mobile phone, and the differentiated interpretations of the mobile phone within Chinese society.

The data used in this chapter were collected by qualitative research, which was conducted in the cities of Beijing and Dongguan from 2003 to 2005. There were 78 informants interviewed in this research, including 59 migrant workers (38 males and 21 females), 10 college students (seven males and three females) and nine white-collar workers (three males and six females). The age of these informants ranged from 18 to 36. Most migrant workers in this study came from three factories in Dongguan: two golf-equipment factories in Tangxia town and a garment factory in Humen town. College students were mainly from Peking University, Tsinghua University, China Youth University for Political Science, and Beihang University. White-collar workers were selected from several international companies in Beijing through the personal connections of the researchers. All the interviews were semi-structured in that they were guided by a set of key questions about informants' mobile phone usage, such as purchase motive and consumption pattern. All interviews were tape-recorded and fully transcribed.

The use of mobile phones in Chinese society

Situation-specific and relationship-specific use

The impact of mobile phone usage on the public/private sphere is significant in most Western societies. Some researchers (Kopomaa 2000; De Gournay 2002; Fortunati 2002; Puro 2002; Rule 2002) hold the view that mobile phones provide an opportunity for people to engage in their private communication in public. Real-time communication draws the attention of mobile phone users away from the public sphere in which they are physically located and into their private domains. This means that the use of mobile phones has created pockets in the public sphere which extend the private sphere at the expense of the public sphere (Fortunati 2002; Puro 2002; Caporael and Xie 2003). Some scholars (Kopomaa

2000; De Gournay 2002) observe that the incursion brought by mobile phone usage is not unidirectional: business affairs and the public sphere also invade private and intimate spaces, such as family, through the mobile phone. To some extent, mobile phone users cannot enjoy a peaceful family life or secure their privacy unless they turn off their mobile phones. Therefore, the incursion of the public sphere into people's private lives is another aspect of the impact of mobile phone usage on the division of the public and private sphere.

However, Chinese users are seldom aware of the public/private division when they use their mobile phones and seldom regard this as a serious problem, even though the use of mobile phones in public places does affect others. Chinese mobile phone users often talk loudly in public spaces, or answer incoming calls regardless of the presence of others. A study compared the mobile phone usage of Chinese and American subjects and found that Chinese participants were prepared to accept business calls during non-working hours, whereas most American participants treated business calls at that time as unacceptable (Caporael and Xie 2003). In order to control their use, and to separate public time from private time, American participants used the screening facilities provided by their phones, or simply turned off their mobile phones when they were not working. In contrast, Chinese users only used the screening facilities or turned off their mobile phones during sleeping hours (ibid.). From the above comparison, it becomes apparent that Chinese participants did not have a clear sense of the division between the public and private sphere, and on most occasions they ignored the facilities provided by mobile phone technology to protect privacy.

The qualitative research conducted by our research team in Beijing and Dongguan found evidence which supported these conclusions by Caporael and Xie. Over 80 per cent of our informants reported that they usually answered incoming calls on their mobile phones in most public places, even if they were in the presence of a third person or a stranger. Few of them regarded this as inappropriate behaviour. This begs the questions: Do these comparisons imply that Chinese subscribers are impolite in their use of mobile phones? Are they prepared to ignore social norms in their use of mobile phones?

In order to interpret this behaviour, we draw on the story of a white-collar worker who had lived in Europe for two years. R was a young white-collar worker who obtained a position in an international company in Beijing. She already had a mobile phone which she had first acquired in 1997 when she was still a college student. Since that time, the device had become indispensable to her. She used it to send over 300 short messages (SMS) every month. She described herself as 'a crazy mobile phone user' who enjoyed chatting with friends through her phone's SMS facility. In 2002 she went to Britain to read for an MA degree, which she completed in 2004. These two years in Britain made a lasting impression on her use of the mobile phone.

> I had the following experience. An American friend was not happy with my mobile phone habits. In Europe, when traveling on public transportation, or

> any other public space, nobody will speak out loudly … I remembered once when my friend and I took the train and went to London, my boyfriend called me. When I spoke to him in a loud voice and my American friend could hear our conversation, he glared at me and complained: 'Must you talk at this time?' He told me that people around me were all looking at me. Later I noticed that British people do use their mobile phones on the train, but they usually talk in hushed tones. After that I began to pay attention to my tone when I answered the phone call in public places. However, I think Chinese have become accustomed to using loud voices. For instance, my father always answers the phone loudly and this makes me feel uncomfortable. 'Oh! President Zhou! I am in Xiamen now …' he will say very loudly. You don't have to let everybody know where you are, do you? But, I think this is his habit, because Chinese do believe that the louder their voice is, the more respect they show to the person who is on the other end of the phone.
>
> (Chu and Peng 2005)

The different views held by R and her American friend regarding the practice surrounding mobile phone usage in public places have mainly resulted from their different understandings of the division between the public and private sphere (Chu and Peng 2005; Chu 2008). More specifically, Chinese do not relate their mobile phone usage to the division of the public and private sphere. They relate the appropriateness of their mobile phone usage to the kind of relationship that exists between the user and the caller at that particular moment (Chu and Peng 2005). In other words, Chinese usually do not use the public/private division as the norm to regulate their mobile phone usage behaviour. Instead, they have a different set of rules. The first relates to attitudes to the boundary between acquaintances and strangers. Chinese mobile phone users regard friends and relatives on the other end of the mobile phone as more important than the strangers around them in a public space. Compared with the possibility of disturbing strangers, the more serious discourtesy for a Chinese person is to speak in a low and dispassionate tone, which would not show enough respect to significant others on the other end of the phone (ibid.). Just as R mentioned, the loud voice on the phone was not only meant to promote clarity in communication, but also to express a degree of respect for those significant others. This mobile phone behaviour does not mean that Chinese people ignore the situation they find themselves in when they use a mobile phone. Another two cases will further enhance this understanding of Chinese mobile phone usage in public spheres. The following conversation took place between the authors and a white-collar worker in an international company.

> I (Interviewer): 'Will you answer mobile phone calls when you are attending a meeting?'
>
> N (Informant N): 'It depends on what kind of meeting I am having. If the meeting is an informal one, I will say "excuse me" to others and answer the call. If it is a relatively formal one, then I won't answer the call.'

I: 'But you will still not turn off your mobile phone in a formal meeting then?'

N: 'We normally do not turn off mobile phones during a meeting. Some colleagues even answer phone calls at the meeting. Sometimes, when the boss is speaking while someone's mobile phone starts ringing, they either go out to answer the call or just answer it in a low voice.'

(Chu and Peng 2005)

In this case, mobile phone usage amongst white-collar workers was situation-specific. Their mobile phone usage did not depend on whether they were in a public sphere or not, but depended upon the type of situation they were in. When calls are received during office hours or in meeting rooms, white-collar workers make an immediate assessment of their situation. If the meeting they are attending is very important, or there are important clients at the meeting, they define the situation as formal or serious. In formal situations, answering a private call is regarded as impolite. When the situation is defined as informal, answering an incoming call is acceptable. Therefore, a boundary is drawn between formal and informal situations, and not between public spheres and private spheres.

Another case comes from the field observation of three migrant workers, X, Y and Z, during a dinner. X, Y and Z were young migrant workers from Hunan province, a southern province in Mainland China. In order to experience an urban lifestyle, they went to Dongguan in Guangdong province. All of them worked in an electronic device factory which had been financed by a Taiwanese entrepreneur. Like other migrant workers, they bought mobile phones as soon as possible after they reached Dongguan.

One night in July 2004, we invited them to dinner and conducted an interview with them during the dinner. Of these three migrant workers, Y is an old friend and has been interviewed by our research team several times. The other two informants, X and Z, were new to us. X is Y's cousin and Z is Y's friend. In the factory where they all work, Y is the team leader of the other two. During the dinner, we noticed that all of them received incoming calls. However, their reactions were very different: Y said 'excuse me' to us and went outside to answer his call, while X and Z immediately turned off their mobile phones without answering their calls. We reassured them that we did not object to them answering their phones but they replied that the calls probably came from friends and were therefore not considered to be important.

In this case, the rule of mobile phone usage was relationship-specific. When and how to answer an incoming call, or reply to a short message, did not only depend on the relationship between the user and the persons present, but also on the assessment of the relative social status between the users and their counterparts. In the dinner situation, the three migrant workers used their mobile phones according to two different relationships: one was their relationship with the researchers; the other was their relationship with each other. Y treated the researchers as his close friends and he believed that using his mobile phone freely in front of his friends was a way of showing them how intimate their friendship was. As the team leader of the other two migrant workers in the factory, he did not care about the reaction of his subordinates when he used his mobile phone. Therefore, he felt free to answer his call during the dinner. X and Z met the researchers for the first time and defined the researchers as friends of their superior. In Chinese society, social individuals treat the seniors or those who have higher social status with respect. In order to show their respect to their team leader and his friends, they turned off their mobile phones or did not answer their incoming calls during the dinner.

Therefore, the public/private sphere division is not of vital concern to mobile phone users in Chinese society. What Chinese users care about is the appropriateness of their behaviour regarding mobile phone usage on relationship-specific or situation-specific occasions. For Chinese users, the use of a mobile phone in a public place is determined by comparing different social situations and differentiated social relationships. First, they consider whether the situation is formal or not. Second, they rank all people involved in the situation, including those on the other end of their phones, in terms of their social status. They place senior persons or those who occupy the upper echelons as the most important ones followed by their intimate friends, acquaintances and then strangers. To some extent, it follows the logic of *chaxugeju* (a differentiated mode of association) in Chinese society (Fei 1991). In his book *Xiangtu Zhongguo*, Fei highlighted the difference between Chinese and Western societies. As he pointed out, individuals in Western societies establish their society by adopting *tuantigeju* (an organizational mode of association), which is embedded in the Christian tradition. In Chinese society, people organize their society in terms of *chaxugeju*, which is based on Confucian ethics. *Chaxugeju* is a metaphor for the ripples that appear on the surface of a lake when a rock is thrown into it. Every individual in Chinese society is the centre and has different relationships with others. These relationships constitute the ripples. The closer the ripple to the centre, the more important the relationship is. Family members constitute the closest ripple to the centre, followed by relatives, friends, acquaintances and strangers. Thus, how a Chinese person treats another one depends on his or her relationship with that person. Therefore, when Chinese use a mobile phone, it is reasonable for them to care more about the feeling of their friends and relatives on the other end of the phone than with that of the strangers in their presence (Chu and Peng 2005). Chinese mobile phone usage in public places is guided by Chinese culture and social norms, which are different from conceptions about the division of public and private spheres in Western societies.

Face and the differentiated social meanings of mobile phone usage

Some studies have discussed the symbolic meaning of mobile phone usage in modern society (Fortunati 2002; Kim 2002; Ling and Yttri 2002; Robbins and Turner 2002; Skog 2002; Katz *et al.* 2003; Campbell 2008; Portus 2008). Some of them (Fortunati 2002; Kim 2002) argue that, at the beginning of its spread, mobile phone usage represented a high-income or success in business. With the penetration of mobile phones into every level of our societies, people gradually related the adoption pattern of mobile phone usage to the public image of the user (Robbins and Turner 2002; Ling 2003; Campbell 2008). The size, colour and other style elements of a mobile phone are very important to the user, since the physical appearance of a mobile phone is vital to the user's self-presentation (Ling 2003; Katz 2006; Campbell 2008). To some extent, the use of an appropriate model or style of mobile phone signifies one's taste or an appreciation of fashion. Besides these aspects of the social meaning of a mobile phone purchase, our qualitative research revealed another important point: the relationship between face and the purchase of a mobile phone in Chinese society (Chu and Peng 2005; Law and Peng 2007; Chu 2008). This relationship is mainly expressed in mobile phone consumption amongst migrant workers.

Migrant workers in Dongguan mainly come from less-developed or poor provinces such as Hunan, Hubei, Sichuan and Jiangxi. Most of them are young boys or girls whose ages range between 18 and 25. Most migrant workers are employed in skilled or semi-skilled jobs in labour-intensive factories. Compared with local residents and formal workers in state-owned enterprises, they are amongst the lowest income groups in the region. For instance, in 2004, the average monthly salary of migrant workers in the Pearl River Delta was about 964.8 yuan (US$147.6) (including overtime pay), which equalled 57.9 per cent of an average urban worker's salary in Guangdong province (Guangdong Statistics Bureau 2005). In spite of their lower incomes, the number of mobile phone subscribers among migrant workers has increased sharply in recent years. In 2002, an insignificant number of high-income migrant workers owned a mobile phone. However, by the middle of 2003, migrant workers had become prolific mobile phone subscribers in Guangdong province. In 2004, the average price of a mobile phone in Dongguan ranged from 1,500 yuan (US$229.5) to 2,000 yuan (US$306), which is a hefty price for most migrant workers. Yet, more and more migrant workers joined the group of mobile phone subscribers. According to a survey conducted by Peking University in the Pearl River Delta, by 2006 over 64 per cent of migrant workers had purchased a mobile phone (Peking University 2006). Moreover, some of them even purchased the latest model with a camera function which cost more than 3,500 yuan (US$535.5) at that time. A supervisor in a golf equipment factory, Chen, gave a vivid description of mobile phone consumption amongst migrant workers in his factory:

> Basically, all of our workers bought mobile phones within these two years. A boy in our factory even spent 3,500 yuan for a new mobile phone. During their leisure time, they enjoy comparing their mobile phones with each other. If one has bought a new mobile phone, he will be very glad to show it to his friends. I don't think it is necessary for them to get mobile phones. They always carry their mobile phones; it seems that having a mobile phone is equal to having an identity.
>
> (Chu and Peng 2005)

This begs another question: why do migrant workers spend so much money on their mobile phones? The manager of another golf equipment factory, Li, gave us an explanation:

> Boys today will not be respected if they do not have a mobile phone in hand. The more advanced the model, the more 'face' (*mianzi*) you gain … in the past people's status was proved by their suit, now it is the mobile phone that counts. The younger the worker, the stronger this desire is … If you are asked for a mobile phone number and you cannot offer one, you will lose face. Presumably, you will have more face if you own the latest model.
>
> (Chu and Peng 2005)

A migrant worker, Wang, also held a similar view:

> If you don't have a mobile phone, you will lose face. That is a great pressure, as you will have a sense of inferiority. The guys in my circle of friends change their mobile phones very frequently. Some of them even buy the latest and most expensive ones.
>
> (Chu and Peng 2005)

It is apparent that purchasing a mobile phone is a way to demonstrate one's status. Face (*mianzi*) is an important and complicated Chinese concept. In her classic discussion on face, Hu (1944) defines face as a kind of social position or prestige, which is achieved by one's successful performance of some social roles recognized by others. Within a particular social network, having face not only means having a relatively higher position, but also merits being respected by others. Among migrant workers, the pressure of owning a mobile phone, to some extent, comes from the need to demonstrate status and prestige (Chu and Peng 2005).

However, these attitudes toward mobile phones are not borne out by this research conducted amongst college students. Most of those interviewed contended that having a mobile phone is not directly related to face. Some college students even went further and said that not having a mobile phone at all was a sign of individuality.

> I (Interviewer): 'Is it shameful to be without a mobile phone?'
>
> A (Informant A): 'No. It is very unconventional (laugh) … I have a classmate who did not have a mobile phone, but he has since bought one for the sake of convenience. If he goes without a mobile phone, he must use other alternatives.'
>
> I: 'Do you agree that the mobile phone is not something we can use to demonstrate our status?'
>
> A: 'Perhaps this was so previously but it is no longer so. However, if the phone has a fashionable design, then it is still a cause for esteem.'
>
> (Chu and Peng 2005)

Does this mean that face is not important to college students? The answer is definitely no: college students do not need to rely solely on mobile phones for their communication because they have other alternatives, such as computers or access to the Internet, which can also demonstrate their status and their ability to use modern technology (Chu and Peng 2005; Chu 2008). Most college students and white-collar workers have their own computers and access to the Internet. Computers and the Internet appear to occupy a position that is at least as important as, if not more than, a mobile phone in daily life (Chu and Peng 2005). One college student believed the functions of a mobile phone were surpassed by computers and the Internet.

> I: 'The development of the mobile phone in activities such as surfing the net, or personal file assistance, is so fast that it seems we are lagging behind.'
>
> A: 'Yes. But, those functions are not necessary for us students. We can easily find a substitute. We can realize those functions on computers. It is only for the sake of convenience that people like to integrate so many functions in a mobile phone.'
>
> (Chu and Peng 2005)

For college students, mobile phones are just one communication tool. Most college students mentioned that their mobile phones were purchased to make their work and lives more convenient. As they said, they had many choices. In contemporary Chinese society, migrant workers and college students belong to different social groups, although they are almost in the same cohort (Chu and Peng 2005). Most college students are from cities and their families are relatively wealthier than families of migrant workers. With a better educational background and with more professional knowledge, college students possess higher social status and have more opportunities for upward mobility than migrant workers have. Generally speaking, college students are regarded as social elites with promising futures. Therefore, they are able to demonstrate face and social prestige in various ways, such as owning a personal computer, and having access to the Internet. Even their college student identity gives them a higher position in the social hierarchy. It is not necessary for them to prove their success by owning an expensive mobile phone.

However, for migrant workers, mobile phone consumption is the only way to show face and prowess with modern communication technology. Few migrant workers can afford to have a personal computer, let alone gain access to the Internet in their homes or dormitories. Furthermore, migrant workers belong to a highly mobile social group (Law and Peng 2006). They not only move between their home villages and destination cities, but also frequently move among different factories and different townships. It is unwise for them to purchase personal computers when they have to change their living places so frequently. More importantly, under China's current household registration system (*Huji Zhidu*), migrant workers are permitted to work in cities but may not be accepted as permanent city residents. Irrespective of how long they have worked in a city, they are always treated as a 'floating population' by local residents and discriminated against (Chan and Zhang 1999; Wang 2001). The social status of migrant workers is much lower than that of college students and white-collar workers. Migrant workers are unable to show face through their career or their identity. Under these circumstances, purchasing an expensive mobile phone becomes an important way for migrant workers to show face. Interestingly, migrant workers seem not to compare their mobile phone consumption with that of their employers. For instance, when a male migrant worker, Liu, who had bought a new mobile phone which cost more than 2,000 yuan (US$306), was told that his employer had bought a mobile phone which only cost 500 yuan (US$76.5), he did not experience any face from this comparison. He explained: 'We are different people. You know, they are the bosses. Even though his mobile phone is much cheaper than mine, he is much richer than me' (Chu and Peng 2005).

Getting face from mobile phone consumption is limited amongst migrant workers. Only when the mobile phone is considered to be a luxurious consumption within a social group, such as migrant workers, can it become a way to show face. When a mobile phone becomes a commonplace device in people's daily lives, it loses this symbolic meaning. To some extent, the different interpretations of the relationship between a mobile phone and the face concept reflect the influence of social status on the symbolic meaning of mobile phone consumption. Although previous researchers in Western society have recognized demographic differences in mobile phone usage, they have seldom explored the distinctive interpretations of the mobile phone among different social groups in a society. Through comparing the different interpretations of mobile phone consumption of various social groups in Mainland China, it can be concluded that the symbolic meaning of the mobile phone is not homogeneous in a society. Different social groups in a social hierarchy will construct the social meaning of mobile phone consumption in different ways.

Conclusion

The mobile phone, as a communication device which is widely used in our society, is more than a technical innovation or a social fashion (Ling 2004). Not only has it had a significant influence on our social activities, but also on our society and our cultural norms, which have in turn influenced our use of mobile phones. From 1987 (when the first mobile phone was introduced into China) to the present day, the use of the mobile phone has rapidly spread across Chinese society (Wang and Cheng in this volume, Chapter 5). Its adoption has taken different forms, being regarded by some as a luxury and by others as a commonplace device needed for everyday existence. Some regard it as an advanced technical innovation and others as an ordinary communication tool. Chinese users as a whole have assimilated mobile phones into their routine work and their lives. During this process, social context and Chinese culture have shaped the social image and symbolic meaning attached to mobile phone usage. Meanwhile,

Chinese users have formed habits, attitudes and behaviour surrounding the use of the mobile phone that are unique to their culture. This does not mean that Chinese users do not share some aspects of their use of the mobile phone with their counterparts in Western societies. What needs to be emphasized is that there is an indigenous perspective with its own social context which is uniquely affected by a global phenomenon: mobile phone communication.

Note

1 This chapter is based on qualitative research conducted in the cities of Beijing and Dongguan from 2003 to 2005. Besides the authors, the research collaborators include Pui-lam Law from the Department of Applied Social Sciences of the Hong Kong Polytechnic University, Shanhua Yang and Weimin Cheng from the Department of Sociology, Peking University, Xiaojing Liu from the Rural Research Development Institute of the Chinese Academy of Social Sciences, and some postgraduate students from the Department of Sociology, Peking University. An early paper (Chu and Peng 2005) was presented at the conference Mobile Communication and Asian Modernities II, Beijing, October 2005.

References

Campbell, S. (2008) 'Mobile technology and the body: apparatgeist, fashion, and function', in J. Katz (ed.) *Handbook of Mobile Communication Studies*, Cambridge, MA: The MIT Press.

Caporael, L. and Xie, B. (2003) 'Breaking time and place: mobile technologies and reconstituted identities', in J. Katz (ed.) *Machines that Become Us: the social context of personal communication technology*, New Brunswick, NJ: Transaction Publishers.

Chan, K.W. and Zhang, L. (1999) 'The hukou system and rural–urban migration in China: processes and changes', *The China Quarterly*, 160: 818–55.

Chu, W.C.R. (2008) 'The dynamics of cyber China: the characteristics of Chinese ICT use', *Knowledge, Technology and Policy*, 21: 29–35.

Chu, W.C.R. and Peng, Y. (2005) 'The use of mobile phones with Chinese characteristics', paper presented at the Conference of Mobile Communication and Asian Modernities II, Beijing, October.

De Gournay, C. (2002) 'Pretense of intimacy in France', in J. Katz and M. Aakhus (eds) *Perpetual Contact: mobile communication, private talk, public performance*, Cambridge: Cambridge University Press.

Fei, X. (1991) *Xiangtu zhongguo* [From the Soil: the foundations of Chinese society], Hong Kong: Joint Publishing.

Fortunati, L. (2002) 'Italy: stereotypes, true and false', in J. Katz and M. Aakhus (eds) *Perpetual Contact: mobile communication, private talk, public performance*, Cambridge: Cambridge University Press.

Goggin, G. (2008) 'Cultural studies of mobile communication', in J. Katz (ed.) *Handbook of Mobile Communication Studies*, Cambridge, MA: The MIT Press.

Guangdong Statistics Bureau (2005) *The Report on Corporation Recruitment in the Pearl River Delta* [Zhusanjiao qiye zhaopin baogao], Guangdong Statistic Information Web. No longer available. www.gdstats.gov.cn/tjnj/table/gaishu_c.htm (viewed 25 March 2006).

Haddon, L. (2003) 'Domestication and mobile telephony', in J.E. Katz (ed.) *Machines that Become Us: the social context of personal communication technology*, New Brunswick, NJ: Transaction Publishers.

Hu, H. (1944) 'The Chinese concept of "face"', *American Anthropologist*, 46: 45–64.

Katz, J. (2006) *Magic in the Air: mobile communication and the transformation of social life*, New Brunswick, NJ: Transaction Publishers.

—— (2008) 'Introduction', in J. Katz (ed.) *Handbook of Mobile Communication Studies*, Cambridge, MA: The MIT Press.

Katz, J. and Aakhus, M. (eds) (2002) *Perpetual Contact: mobile communication, private talk, public performance*, Cambridge: Cambridge University Press.

Katz, J., Aakhus, M., Kim, H. and Turner, M. (2003) 'Cross-cultural comparisons of ICTs', in L. Fortunati, J. Katz and R. Riccini (eds) *Mediating the Human Body: technology, communication and fashion*, Mahwah, NJ: Lawrence Erlbaum Associates.

Kavoori, A. and Chadha, K. (2006) 'The cell phone as a cultural technology: lessons from the Indian case', in A. Kavoori and N. Arceneaux (eds) *The Cell Phone Reader: essays in social transformation*, New York: Peter Lang.

Kim, S. (2002) 'Korea: personal meanings', in J. Katz and M. Aakhus (eds) *Perpetual Contact: mobile communication, private talk, public performance*, Cambridge: Cambridge University Press.

Kopomaa, T. (2000) *The City in Your Pocket: birth of the mobile information society*, Helsinki: Gaudeamus Kirja.

Law, P.L. and Peng, Y. (2006) 'The use of mobile phones amongst migrant workers in southern China', in P.L. Law, L. Fortunati and S. Yang (eds) *New Technologies in Global Societies*, Hackensack, NJ: World Scientific.

—— (2007) 'Cell phones and the social lives of migrant workers in southern China', in R. Pertierra (ed.) *The Social Construction and Usage of Communication Technologies: Asian and European experiences*, Diliman, Quezon City: The University of the Philippines Press.

Leonardi, P., Leonardi, M. and Hudson, E. (2006) 'Culture, organization, and contradiction in the social construction of technology: adoption and use of the cell phone across three cultures', in A. Kavoori and N. Arceneaux (eds) *The Cell Phone Reader: essays in social transformation*, New York: Peter Lang.

Ling, R. (2003) 'Fashion and vulgarity in the adoption of the mobile telephone among teens in Norway', in L. Fortunati, J. Katz and R. Riccini (eds) *Mediating the Human Body: technology, communication and fashion*, Mahwah, NJ: Lawrence Erlbaum Associates.

—— (2004) *The Mobile Connection: the cell phone's impact on society*, San Francisco, CA: Morgan Kaufmann.

Ling, R. and Yttri, B. (2002) 'Hyper-coordination via mobile phones in Norway', in J. Katz and M. Aakhus (eds) *Perpetual Contact: mobile communication, private talk, public performance*, Cambridge: Cambridge University Press.

Mesch, G. and Talmud, I. (2008) 'Cultural differences in communication technology use: adolescent Jews and Arabs in Israel', in J. Katz (ed.) *Handbook of Mobile Communication Studies*, Cambridge, MA: The MIT Press.

MIIT [Ministry of Industry and Information Technology] (2004) *2004 nian 12 yue tongxin hangye tongji yuebao* [Monthly Statistical Report on the Information and Communication Industry: December 2004]. Online. Available: www.miit.gov.cn/n11293472/n11293832/n11294132/n11302706/11766170.html (accessed 24 January 2009).

—— (2011) *2010 nian quanguo dianxinye tongji gongbao* [The Report on National Information and Communication Industry in 2010]. Online. Available: www.miit.gov.cn/n11293472/n11293832/n11294132/n12858447/13578942.html (accessed 12 April 2011).

Peking University (2006) *Zhusanjiao nongmingong jiqi liudong qingkuang baogao* [The Report on Migrant Workers and their Mobility Status in the Pearl River Delta], unpublished report.

Portus, L. (2008) 'How the urban poor acquire and give meaning to the mobile phone', in J. Katz (ed.) *Handbook of Mobile Communication Studies*, Cambridge, MA: The MIT Press.

Puro, J. (2002) 'Finland: a mobile culture', in J. Katz and M. Aakhus (eds) *Perpetual Contact: mobile communication, private talk, public performance*, Cambridge: Cambridge University Press.

Robbins, K. and Turner, M. (2002) 'United States: popular, pragmatic and problematic', in J. Katz and M. Aakhus (eds) *Perpetual Contact: mobile communication, private talk, public performance*, Cambridge: Cambridge University Press.

Rule, J. (2002) 'From mass society to perpetual contact: models of communication technologies in social context', in J. Katz and M. Aakhus (eds) *Perpetual Contact: mobile communication, private talk, public performance*, Cambridge: Cambridge University Press.

Skog, B. (2002) 'Mobiles and the Norwegian teen: identity, gender and class', in J. Katz and M. Aakhus (eds) *Perpetual Contact: mobile communication, private talk, public performance*, Cambridge: Cambridge University Press.

Wang, C. (2001) 'Xinshengdai liudong renkou de shehui renting yu chengxiang ronghe de guanxi' [The social identification of the new generation of migrant workers and its relationship with city–village merger], *Shehuixue yinjiu* [Sociological Research], 93: 63–76.

14 Privatizing public spaces and personalizing private spaces

The challenge of mobile communication in Beijing[1]

Boxu Yang, with Bo Gai and Li Li

Introduction

This chapter concerns the role of the mobile phone in today's social structures in Beijing. China is in a unique historical movement. According to the data provided by the Ministry of Industry and Information Technology (MIIT) of the PRC, Beijing is not only at the heart of the movement but also in the leading position in terms of Internet penetration (69.4 per cent) and mobile phone adoption (121.4 per cent) in China (MIIT 2011). New media communication has become part of the daily lives of Beijing residents. Against this backdrop, it is interesting to know how social relationships are changed in Beijing in an age of mobile communication. In the West, we are told how the new media communication resulted in the emergence of 'networked individualism' and 'imagined privacy' in human experience, but how does the Chinese context read this epoch? In order to answer the question, the focus is on analysing what kind of social ties are strengthened or weakened by the spatial change brought about by the diffusion and use of the mobile phone. Such enquiries have rarely been explored in the urban Chinese context. As such, this is going to have significant research value for those who would not want to overlook the Chineseness of cyber China.

Review of theoretical issues and relevant empirical findings

The diffusion of new media such as the Internet and the mobile phone has accelerated the emergence of 'networked individualism' and 'imagined privacy' in the context of globalization. It is important to review the content of these two concepts, as well as relevant empirical findings surrounding them, before we can depict the similarity and difference of experiencing new media communication between the West and China.

Networked individualism: the change of social structures

The pre-industrial community, spatially compact and densely knit (Barthélemy and Contamine 1985; Thébert 1985; Ward 1999), had door-to-door connectivity (Wellman 2001, 2002). Modern technologies such as the automobile and telephone brought change to a place-to-place community that is no longer limited to the neighbourhood (Wellman 1988, 1999, 2001, 2002; Wellman and Leighton 1979).

New media such as the Internet and the mobile phone herald a new transition into person-to-person connectivity. It is assumed that 'the shift to a personalized, wireless world affords "truly personal communities"' (Wellman 2001, p. 238). In such a context, people are experiencing a transition to the rise of networked individualism (Castells 2000; Haythornthwaite and Wellman 2002; Wellman 2001, 2002).

The Internet and the mobile phone are inseparable or complementary as far as their roles in forming networked individualism are concerned. Some researchers specifically study the relationship between on-line activities and the formation of social ties. The amount of time spent on the Internet is expected to be closely related with devotion to community ties. It is assumed that the more people are engaged in affairs on-line, the less they interact with their neighbours (Coffey and Stipp 1997; Robinson and Godbey 1997). Since community should not be treated as a 'zero-sum game' (Matei and Ball-Rokeach 2002; Wellman and Gulia 1999), some scholars emphasize that the Internet supports a variety of community ties, not only weak ties but also strong and intimate ties (Wellman and Gulia 1999), and strong ties had already spread beyond the neighbourhood before the introduction of new media such as the Internet and the mobile phone (Wellman 1979; Wellman and Tindall 1993; Wellman and Wortley 1990). Recently, the strength and nature of the social network tie prior to the emergence of mediated communication has been taken into consideration to synthesize these disparate findings (Haythornthwaite 2001).

From this perspective, the Internet is interpreted as having no negative impact either on an overall sense of community (Quan-Haase *et al.* 2002) or on communication with intimate ties (Boneva and Kraut 2002; Howard *et al.* 2002; Katz and Rice 2002). Moreover, a close relationship tends to increase the frequency of media use (Chen *et al.* 2002; Katz and Aspden 1997; Quan-Haase *et al.* 2002; Robinson *et al.* 2002), and individuals with a stronger feeling of belonging to a neighbourhood are more likely to make a friend on-line (Hampton and Wellman 2002; Matei and Ball-Rokeach 2002).

But the usage of the Internet is usually in fixed locations in Beijing. That is, the usage of the Internet may be in a private location, but not necessarily in a personal space. It has been reported that one of the key differences between the mobile Internet and the PC Internet is that high-intensity PC Internet users tend to spend less time with intimate ties, while heavy mobile Internet users are more devoted to socializing and interpersonal communications (Castells *et al.* 2004; Ishii 2004). Such findings have great implications for mobile phone research.

Some studies have demonstrated that mobile phone users are more sociable than non-users, and that the use of e-mail via the mobile phone enhances sociability (Hashimoto *et al.* 2000; Tsuji and Mikami 2001). Furthermore, the mobile phone is considered as going further in affording person-to-person connection than the personal computer (Wellman 2001). The mobile phone increases individual selectivity in choosing a relationship (Fortunati 2002a; Matsuda 2005), and it in a sense maintains and reinforces the established ties (Dobashi 2005; Fortunati 2000; Ling 2000). There is a positive view of the impact of the mobile phone on face-to-face communication as well. The emergence of a wireless networked society makes the informal, geographical co-ordination of intimate small groups, or *rendezvous*, achieve its full sense (Castells *et al.* 2004). Clearly, individuals, with the assistance of the mobile phone, are regarded as being able to arrange face-to-face activities with family, friends, and workmates at any time.

While the literature reviewed above shows different findings on social ties, it is clear that the social relationships are changing with the assistance of the Internet and mobile phone. We are concerned with the changes as well as the direction of the changes in respect to mobile phone users in Beijing.

Imagined privacy: the public and the personal

Another unique feature of mobile phone communication is its privacy or imagined privacy. With the diffusion of mobile media, especially the mobile phone, the interplay of the public and the private has become another main concern. In this context, the concept of 'imagined privacy' comes into being. Many researchers have discussed the mobile phone's role in blurring the boundary between the public and the private by paying attention to the use of the mobile phone in public places. Before the introduction of the mobile phone, there were certain 'public manners' regulating what and how to talk about private matters in public spaces (Castells *et al.* 2004; Kim 2002). However, 'these manners seem suddenly to have evaporated in this era of perpetual contact' (Kim 2002: 65).

The 'imagined privacy' is a term used in the social context where the conception of space and time has been converted with the assistance of modern communication technologies. That is, with the emptying of time, the space is torn away from the place by making connections with absent others (Giddens 1990). For instance, due to its mobility and portability, the mobile phone provides people more freedom in maintaining a 'perpetual connection' despite the material and physical surroundings (Gergen 2002). Before the adoption of the mobile phone, the use of time was generally defined by location (Fortunati 2002b). While afterwards, communication abstracted from location by the mobile communication stimulates the dissolution of the mapping between activities and places, and every place seems to be 'for everything all the time' (Agre 2001). As some scholars have pointed out, the mobile phone removes the need to stay by the telephone and fundamentally liberates individuals from place (Wellman 2001), and thus reduces the 'place-centeredness of schedules' (Palen *et al.* 2001).

According to Fortunati, the sense of belonging to a place 'is actually transformed into the sense of belonging to one's communicative network' (Fortunati 2000: 9). Moreover, she points out that the mobile phone not only enables people to 'capture the intimacy of interpersonal relations … in a public dimension' but also opens up 'the possibility of conquering new intimate communicative spaces at a personal level in the home' (Fortunati 2002a: 49–50). This is not only a sign of the rise of networked individualism but also implies the division of space in the mobile society. As Gergen (2002) puts it, the mobile phone conversation establishes an 'inner space' where people are communicating with each other as well as an 'outside space' where the significance of the present outsiders is reduced to a minimum. This phenomenon has also been described in other terms such as the physical space and the conversational space. As Palen *et al.* (2001: 121) indicate, 'When mobile phone users are on the phone, they are simultaneously in two spaces: the space they physically occupy, and the virtual space of the conversation.' In their research on mobile email use, Ito and Okabe conclude that mobile phones 'create new kinds of bounded places that merge the infrastructures of geography and technology, as well as techno-social practices that merge technical standards and social norms' (Ito and Okabe 2005: 260).

Scholars adopt terms such as 'full-time intimacy' (Matsuda 2005), 'telecocoon' (Habuchi 2005), or 'co-presence' (Ito and Okabe 2005) to describe the mobile phone's role in maintaining intimate ties. The term 'imagined privacy' puts more emphasis on the interplay between the mobile phone user and the surroundings. In public places, mobile phone users tend to behave differently from non-users, and are more likely to avoid eye contact with people around, often ignoring them (Castells *et al.* 2004; Murtagh 2002). Thus, mobile phone users absorbed in a mobile phone conversation in a public place are considered to be ceasing to be 'full participants in the immediate context' (Gergen 2002: 239).

But the concept of imagined privacy primarily emerged in facing the challenge posed by mobile phone communication in public. In other words, the concept mainly concerns privatizing public spaces by means of imagination. It does not deal with social situations such as personalizing private spaces (at home by means of SMS (Short Message Service), for instance). Indeed, our research addresses the question and assumes that networked individualism, privatizing public spaces, and personalizing private spaces are interrelated because of the diffusion and implementation of mobile phones in Beijing.

Research methods

This study is exploratory in nature. The data were collected by conducting in-depth interviews with 18 mobile phone users who are Beijing residents. There are two kinds of residents in Beijing. Those with registered permanent residency enjoy the full social welfare the city provides; those without registered permanent residency do not. This chapter primarily concerns the role of the mobile phone in the social networking of the registered residents. Thus, migrant workers are excluded for they are classified as temporary residents by the Beijing Municipal Government. College students are not considered because we are interested in exploring the impact of mobile phone communication on different kinds of social ties. The social lives of college students are primarily limited to the campus in China. All the interviews were carried out face-to-face and tape recorded with the permission of informants between January 2006 and January 2007.

We used purposive sampling and carefully selected the 18 informants based upon their education, occupation, income, gender, age, and marital status. In so doing, we are able to roughly classify the 18 mobile phone users into three groups: (1) young (aged 21 to 33); (2) middle-aged with higher SES (socioeconomic status); (3) middle-aged with lower SES. While such a sample would not provide a good description of each group as a whole, it might be useful for general comparative

Table 14.1 The socio-demographic characteristics of the informants

Informant	*Age*	*Sex*	*Marital status*	*Occupation*	*Annual income (USD)*	*Education*
A	21	M	S	Clerk in private firm	2,500	Junior college
B	22	F	M	Receptionist in museum	2,800	Junior college
C	26	F	M	Part-timer at home	2,500	Junior college
D	26	F	S	Clerk in private firm	7,750	MA
E	27	F	S	Journal editor	6,750	BA
F	27	F	S	Clerk in foreign firm	9,000	MA
G	28	M	S	Technician	10,625	Trade school
H	31	M	M	Civil servant	6,250	MA
I	32	M	S	Bank accountant	15,650	MA
J	33	M	M	Bicycle repairman	1,000	High school
K	37	F	M	Associate professor	3,750	PhD candidate
L	37	F	M	Manager	6,250	BA
M	38	M	D	Junior high-school teacher	3,125	BA
N	39	M	M	Entrepreneur	62,500	BA
O	43	M	M	Small-restaurant owner	—	High school
P	44	F	D	Bookkeeper	1,875	Junior college
Q	45	M	M	Associate professor	18,750	PhD
R	45	F	M	Waitress	3,125	High school

purposes. Table 14.1 shows the socio-demographic characteristics of the 18 informants.

Individualism and social ties: the role of the mobile phone

By its very nature, the mobile phone implies a capacity to control and manipulate space. Space is closely related to social relations. Will social structures change when the space becomes further disciplined because of the use of the mobile phone? Studies from the West suggest that mobile phone communication is able to increase individual selectivity in choosing relationships and reinforce the established ties in a sense (Dobashi 2005; Fortunati 2000, 2002a; Ling 2000; Matsuda 2005). In this section, we analyse the dynamics of mobile phone communication and social relations in respect to kinship ties, best friends, and other social relationships in Beijing.

Mobile phone communication, kinship ties, and non-traditional love ties

Kinship ties have traditionally been the basic component of social structure in China. This might still be true as far as mobile communications are concerned. Our data show that mobile phone communication between parents and children is almost a daily routine and affective in nature. Our informants whose hometowns are not Beijing call their hometown parents at least once a week. Moreover, it is the mother and the daughter who are more likely to take the initiative and maintain their parent–child ties, which is similar to the findings on the gendered use of the mobile phone in Japan (Dobashi 2005).

The majority of our informants also enjoy frequent mobile phone or Internet communications with their selected relatives. Such strong ties include not only blood-related cousins or sisters or brothers, but also relatives by marriage. For instance, while a 28-year-old technician acknowledges that 'blood is blood, it is different after all', he also meets his cousin's husband for lunch quite often. A 21-year-old male clerk claims that mobile phone communication with his father's sister-in-law has already become something of a routine activity for him: 'I'll send messages to my aunt at each weekend, for we are in a very good relationship. And probably we will make an arrangement for a family party.'

Although almost all the informants who are married have kept frequent contact with their spouses through the mobile phone, they communicate more for clarification or instrumental purposes. One informant, a 37-year-old female associate professor who lived in one of the major southern Chinese cities, had moved to Beijing for her PhD degree. Her husband and child are still in the city she left. She explains:

> My husband and I talk very often on the phone. He usually calls me. If I did not pick up the phone, he got angry. If he accepted my explanation later on, it would be alright. Otherwise, he would be still angry. … If there is no mobile phone, there will be less misunderstanding. I sometimes really don't like the mobile phone. … That is to say, the phone … brings too many misunderstandings, too many restraints. If I did not answer one phone call from him or failed to reply to his SMS right away, I had to spend a lot of time in explaining later, explaining where I went and what I did. I really don't like to use the mobile phone in situations like this.

While the case of the female associate professor may seem extreme because of the physical distance, it is true that the mobile phone is mainly employed by the married informants for the purpose of coordination, such as picking up children from school. Mobile phone communication between informants who are engaged or have steady boyfriends or girlfriends shows similar patterns. One 32-year-old accountant who works for a bank describes his mobile phone communication experience with his fiancée:

> Now we call each other more for practical matters. It was emotional and affective exchanges before when we were boyfriend and girlfriend. Now it is just for talking about things, a matter of communication. The mobile phone is just a tool of communication.

A 27-year-old journal editor discusses how she used the mobile phone to manipulate two intimate relationships during the same period:

> I used QQ [popular Internet chat software in China] to chat with the boy I met after I had a boyfriend. Then, I used the phone to talk to my boyfriend when I was at home. I got to know him in an organized trip to Tibet. The trip members were gathered through the Internet. ... Personally, I don't like phones.

It appears that the mobile phone means different things and is used differently in different stages of intimate relationships. It could be restraining, instrumental, or even manipulative. But for the younger and unmarried informants, it means liberation in terms of disciplining space, controlling time, and reaching out for intimate or affective relationships. Indeed, our informants who are single or divorced are often able to have and maintain intra- or inter-city romantic relationships by means of mobile phone communication, especially SMS. SMS is preferred not only for its low cost in China, but also because it makes it easier to express affective feelings. A 27-year-old female clerk who has a Master's degree and works for a private firm tells us:

> ... I think as Chinese we are more reserved. If you let him express his feelings directly, he may be embarrassed. An SMS message is different because it is not a direct conversation. So it is possible to have a higher degree of freedom on the expression.

Mobile phone users send out and receive huge numbers of SMS messages every day in China. Nevertheless, it is only part of the picture for the development of romantic relations, as the 27-year-old journal editor told us:

> SMS is only part of communication for significant relationship development. I sure have other channels to communicate with intimate male friends. For instance, we can go to the Internet and chat or meet face-to-face. ... Put it this way, if there is such a development in romantic relationships or any sign of moving towards a more intimate direction, it cannot be accomplished by SMS alone. It is only part of the interaction.

The reaching out for a romantic relationship through the mobile phone may mean many things. For the informants who are already married, it indicates dissatisfaction with their spouse. Studies in gender relations and feminist movements show that dissatisfaction from the female side has a long history (Rong 2001; Su 2004). The new media such as the mobile phone undermine the spatial limitation.

China is well known for its wall-and-gate mentality and constructions. Walls and gates, with their many implications, are mainly used to divide spaces and prevent undesired communication. In urban China today, steel gates are installed in every high-rise apartment in the hope of constructing a safe and private space in these times of modernization and globalization. But modernization and globalization, assisted by the new media, are processes in reverse for they are tearing the walls down in a remarkable way both domestically and internationally. For the mobile phone, it helps privatize public spaces or turn family spaces into personal spaces whenever it is necessary. For instance, the silent and personal SMS messages pass through the steel gates and concrete walls. This has great implications for social relations. A divorced 44-year-old bookkeeper who simultaneously maintains a romantic relation and three intimate relations with the opposite sex explains, 'In general, I use the mobile phone to communicate with my personal and intimate friends. … I work in an office. Everybody would know what I was saying if I had used the office phone.'

All three of the intimate male friends of this informant are married men. She uses the fixed phone for same-sex friends for it does not matter if others around will hear her. But the mobile phone is reserved for her personal and intimate male friends because the mobile phone allows her to go to another room or place where there are no other people around.

Another informant, a married 45-year-old male, who works for a university as an associate professor, has had a long-term extramarital affair with a journal editor. He and his mistress meet at the same place every Thursday afternoon. They usually schedule the tryst on MSN (Microsoft Network). Then they use the mobile phone to coordinate further. They also send short messages to one another to check if it is convenient for them to chat on MSN.

Almost all our married informants who are white-collar workers or professionals have or have had extramarital affairs or intimate ties with the opposite sex regardless of gender. It is reported that there was a sharp increase in the number of divorces in 2007. Compared with the statistics in 2006, the number of registered divorces increased 5.9 per cent, while the registered marriages decreased 31.2 per cent (Beijing Municipal Bureau of Statistics 2009). Besides, it is found that one-third of Chinese marriages are viewed as unsatisfying (Ye and Lin 1998). Thus, we have reason to argue that the rising divorce rate and mobile phone communication in urban China imply that it is getting easier for both the divorcees and people who suffer marital problems to seek emotional support from the people they appreciate.

While relationships as such are greatly assisted by the mobile phone, the space for the married informants is redefined by SMS. The messages are regarded as personal. It is acknowledged by virtually all the married informants that they do not check their spouses' short messages, although some of the unmarried informants do secretly check their boyfriends' or girlfriends' short messages.

The married informants on lower incomes use the mobile phone much less often, mainly for instrumental purposes. They seem to adhere to the traditional idea that marriage is not primarily based on emotional attachment (Fei 2006: 51–67). The instrumental needs of the family are more emphasized. The time-consuming extramarital adventures are rare so far as our informants with lower income are concerned.

However, this does not always imply that they have a higher moral standard or dislike modern communication technologies. A 33-year-old self-employed bicycle repairman whose annual income is between 700 and 1,100 US dollars tells us:

> I rarely contact the opposite sex except my wife. … It is not that I am afraid of my wife. I am just not used to it. It is not necessary for I am very busy and trying to make a living. I have little time to go out and have some leisure activities.
>
> … I don't share my problems with others. I take my responsibilities. I don't like to talk [about my problems] to other people. If there is something that really bothers me, I will discuss it with my wife. After all, my wife is the closest to me.

Another informant, a 43-year-old small-restaurant owner who has a modest annual income, behaves in a similar way. In his current marriage, he does not have female friends and sends short messages only to his wife. For him, phone calls are mainly for business purposes. He is trying to make a living for his family. The restaurant and home are the two places he spends his time. It seems that economic constraints and the pressure faced in daily life cause some people to feel that marriage and family may be the most important resources they have during the modernization and globalization processes. That is, family members depend upon each other's support for survival. Similar patterns are reported among migrant workers in China. The migrant workers are mainly from poor rural villages. Law (2005) indicates that the first generation of migrants do not show marriage-related emotional problems even though the couples are physically separated, because the most important thing for them is to fulfil the functional needs of the family (cf. Yuen *et al.* 2004).

Mobile phone communication, best friends, and other social ties

Our informants' experiences suggest that mobile phone communications are not only associated with the changes in intimate lives by redefining personal space but also related to the shift of interaction patterns with best friends and other social ties. As in other places, most of our informants claim that they have two or three best friends. However, the majority of our informants do not meet or communicate with their friends frequently in spite of having a mobile phone.

To our informants, best friends are generally of the same sex. Our two young unmarried male informants neither visit nor call their claimed best friends regularly, although they do sometimes use the phone to affirm their ties. As one of the two told us, '… we call each other once in a while. There is no fixed frequency. A call is made whenever it occurs to me that it is time for me to make the call.' We should point out here that it is neither physical distance nor the long-distance cost of phone calls that prevents them from seeing or calling each other regularly.

The three female informants who are young and single, however, show different interaction patterns with their best friends. They communicate with each other regularly and frequently. A 26-year-old clerk who works for a private firm indicates, '… we are on the MSN anyway. So we "see" each other very often and chat for a few minutes each time. … If something urgent comes up and my best friends are not on MSN, I will use my mobile phone and send them a short message. But there is no urgent matter normally.' A 27-year-old unmarried female professional who works for a foreign firm explains her experiences with her best friend living in another city:

> My best friend and I send SMSs to each other daily. … For instance, she will send me short messages on anything. I cannot say that I know what she is doing on a 24-hour basis. But I pretty much know where she is or how she is doing every day. I will know if there is any change in her life anytime. She would send me SMSs from road, airport, or railway station.

Our middle-aged informants deal with their best friends differently compared with the younger ones in an age of mobile communications. Our informants who are professionals or white-collar workers do not meet their best friends weekly. They meet up occasionally such as on a public holiday or when something comes up. Nor do they call or send short messages to each other very often, regardless of gender. A female project manager puts it this way:

> I think that the friendship among best friends has a stronger base and deeper roots. Yes, it is not necessary to see each other very often for the ties were established in an earlier stage when we were classmates in school. We may bump into each other occasionally during some events, for we work in the same profession. Then we are all very happy and talk about what is going on with our lives.

To the urban Chinese, best friends are usually those friendships forged in school. They are strong ties but not active ties as far as our professional or white-collar worker informants are concerned. Again, the less active communication pattern is not related to physical distance.

But our middle-aged informants who are blue-collar workers or who have lower income tell us differently. Generally speaking, they not only see their best friends often but also communicate with them frequently. The 43-year-old small-restaurant owner says:

> I have seven or eight good friends. We grew up together. I meet two [of them face-to-face] often, not the others. … For instance, when there is something unpleasant between me and a friend, I'd like to discuss it with my best friends.

Another informant, a 45-year-old waitress, tells us that 'best friends are the ones who come to see each other often, I think so'. When she was laid off, her best friends visited her at home to chat rather than simply making a phone call. Now she has a job and often goes for a walk with her friends after work, especially during the summer. Now the waitress is busy with her new job. But she does not forget to call one of her best friends, who lives in another part of the city and is still laid off, regularly to provide her with emotional support.

Before the diffusion of mobile communication technologies, Chinese were more likely to connect with extended social ties, such as classmates, workmates, comrades-in-arms, and neighbours, depending upon their SES and the neighbourhoods where they grew up. We may regard such social networks as traditional networks. One of the properties of traditional networks is that, within a network, the members tend to know each other well. Our findings suggest that such social networks still exist in today's Beijing. The interaction patterns are similar to a large degree regardless of age, gender, and SES.

The younger informants tell us that they have about five or six good friends or extended close ties as explained above. But both the younger males and females only meet their friends on special occasions such as birthdays, marriage ceremonies, or the Chinese New Year, although they sometimes communicate with each other through the Internet and mobile phone. That is, their face-to-face meetings are often group gatherings. As the 27-year-old female who works for a foreign firm indicates: '[The major hurdles are coordination.] The more people are included, the harder it is [to organize]. It is too difficult to coordinate the time [that suits everybody].' Instead, the younger females sometimes use the Internet. For instance, the 27-year-old journal editor sometimes joins group chats with her friends on MSN.

Almost all our middle-aged informants not only meet their good friends in the traditional sense occasionally, but also use the phone to communicate with them more for instrumental purposes. That is, they are more likely to contact their good friends on special occasions such as when they need their friends' material or emotional supports. Indeed, the 45-year-old waitress indicates that the only period during which she visited her old neighbours and classmates often was the period when she was laid off. In her own words, '... many were laid off. It was boring at home.'

The state of being busy, or busyness, which is some informants' excuse for their shortening of time spent with good friends, has become a way of life during the modernization and globalization processes in urban China. Life is busy. But city dwellers are communicating for social purposes without precedent in history both in volume and frequency by means of the mobile phone and the Internet. Our informants' busy lives do not prevent them from seeking new friends and forming social networks that are individualist in nature.

Mobile phones have the capacity to store telephone numbers. Although the size of one's telephone network may be due to one's work or occupation, the number of telephone numbers stored in one's mobile phone implies the range of one's social connectivity at least to some degree. The data from our informants reveal that they store up to and over 400 numbers. All of our informants except those with lower incomes indicate that they keep more than 100 phone numbers in their mobile phone. Age is not closely related to the range. Income is more significant on the matter. Our informants on lower incomes with fewer prospects, such as the waitress, small-restaurant owner, and bicycle repairman, report that they have fewer than 30 phone numbers stored in their mobile phones, and some have none. This correlates with the amount of money they pay for their mobile phone service. The lower income informants spend from 2 to 6 US dollars per month while the others spend from 10 to 60 US dollars for the service. Some of the 'big spenders' have the mobile phone service fees either partially or entirely paid by the firms they work for.

We then asked our informants whom they communicated with besides business associates or workmates, parents/children, and personal and intimate friends. The answer is they make friends based upon hobbies and various other needs. This is particularly true for some of our younger informants who actively seek friends through the Internet. The 28-year-old male technician tells us:

> I have different circles. People within the same circle know each other, but of course they do not know the ones in other circles. For example, I have a singing circle. And I also have a circle within the company I work for. The two circles do not overlap. ... I got to know my singing friends through a website called 'Accompany me to travel'. We started to chat on the net. Then we felt that we could get along with each other. We meet if we feel good. Otherwise, I will just terminate the relationship. Those friends have no personal interests involved . . . just for fun, with no emotions attached.

Maintaining such relationships is less important than joining and developing the circle or individualism-oriented network indicated above, although there are such possibilities. They communicate with and meet each other merely to have a good time together. In such circles, history and future are not important. What is key is the ability to get along for the time being. Our younger female informants show a similar pattern. They also form or develop interests-directed 'circles' through Internet tools such as MSN. Again, their circles do not overlap. That is, there is no linkage between the groups.

While the Internet may be employed to join or form a particular circle of friends, the mobile phone can be used to maintain the personal needs-based new ties. Although both the younger males and females use their mobile phones to communicate with personal needs-based new friends, they make it clear that their friends do not know each other by and large. In other words, such networks often have a star shape. Everyone is in the centre of their network developed by means of the mobile phone. Networks as such often have intercity and even international linkages with the assistance of new media in general and SMS in particular. Again, history and future are not major concerns. These self-centred networks are often quickly developed or dissolved in an age of mobile connectivity. It all depends upon our younger informants' will and needs.

Our middle-aged white-collar or professional informants also reach out and develop relationships beyond their traditional Chinese social ties as mentioned earlier. However, they rarely form the circles indicated by the younger informants. By means of the mobile phone or the Internet, they are more likely to develop and maintain a few relatively stable ties or linkages. That is, one may have a few 'mobile' friends who are not cross-linked. However, the person at the centre does share his or her life history to each tie at least to some degree. With the lower income informants, communications are limited to the traditional network ties regardless of the availability of mobile phone communication.

Referring to the migrant workers from rural villages, Law (2005) reveals that mobile phones not only help them to maintain existing kinship relationships in expanded spatio-temporal contexts, but also to prolong new social relationships. Our findings on Beijing residents show similar communication patterns, but different motives lie behind them. While blood ties are still very strong for our informants in general, the younger ones with higher incomes have started to form hobbies or needs-driven circles and explore multiple intimate relationships with the assistance of the mobile phone. The middle-aged informants who enjoy higher income are greatly encouraged by the mobile phone in terms of having extramarital affairs and developing other social ties beyond kin, workmates, classmates, and comrades-in-arms. Indeed, the essential feature of the mobile phone is its mobility.

Mobile phone communication and spaces

Spaces are closely related to social relations. We act differently in different spaces in terms of our understanding of public space, private space, and personal space. However, the mobile phone allows its Beijing users to reach out and establish ties by privatizing public spaces and personalizing private spaces, such as family. While mobile phone conversations treat public space as private quarters, personal short messages effortlessly penetrate the apartment walls and steel doors that guard each family, invading even the marriage contract in Beijing. In this section, we further demonstrate our points by analysing imagined privacy and other related categories.

Mobile phone communication in public space

Previous research reveals that the use of the mobile phone in public contributes to the blurring of the boundary between the public and the private (Castells *et al.* 2004; Kim 2002), and the mobile phone brings a loss of the tie between place and person (Laurier 2001). Our data show similar results. Mobile phones are commonly used in public spaces by our informants. Users with higher SES regardless of age and gender are more likely to keep their voice low or switch to their hometown dialects whenever private matters are dealt with. The 31-year-old male civil servant reveals his high awareness of privacy: 'I am afraid of being overheard by people around, even strangers, when it comes to my privacy. Sometimes I speak in dialect, but I feel people all over the country can understand the dialect.'

Our informants with higher income and education are more alert when they communicate through the mobile phone in public spaces. But it seems that less educated or lower income users do not carry the privacy burden when in a public space. The 44-year-old female bookkeeper who has an annual income between

1,800 and 1,900 US dollars describes her mobile phone communication experience in public spaces:

> However, in public spaces where there are no acquaintances around, the situation is quite different from that in the office. In the office, you are surrounded by colleagues who care or are even curious about whom you are talking to and what you are talking about on the phone. But in streets or department stores, I answer the phone very often. It doesn't matter, because nobody knows who you are, and I feel liberated when I am talking on the mobile phone.

Almost all of our blue-collar informants claim that nearby strangers do not exert much pressure upon their mobile phone conversations. Anonymity is enough to make them feel that they have all the privacy needed when they speak on the mobile phone in public spaces. For instance, the 45-year-old waitress 'never thought that people on the bus would overhear' when she was making a mobile phone conversation. Indeed, as demonstrated in the previous section, there is not much for them to conceal as far as their social interactions are concerned. Their privatization of public space has little to do with the imagined privacy. They just do it; they have no need to imagine. Indeed, they are more concerned with the cost of each mobile phone call than anything else. Accordingly, their mobile phone conversations are more likely to be brief.

While different groups perceive the privacy issue differently in public spaces, they all agree that restaurants are exceptions. More or less, speaking loudly is tolerable in Beijing restaurants. That is, it is already a privatized place even before the diffusion of the mobile phone to some extent. As the 37-year-old female manager told us, 'It doesn't matter if I use the mobile phone when I am dining in a restaurant. I answer mobile phone calls directly and make calls according to my need.'

To our informants, private mobile phone signals penetrate the walls of office buildings too. Sending or receiving personal short messages seems an accepted behaviour in office meetings and on other business occasions. When privacy is assured because of the silent nature of SMS, the office spaces are personalized to some degree. Moreover, taking mobile phone calls in the office is tolerated. The middle-aged or near-middle-aged informants are able to either temporarily create a personal space by walking to some place where there is no one around or manipulate the office environment by speaking in a different way. For instance, the 32-year-old bank accountant says he will try to 'speak in the usual tone, and not to reveal real feelings or opinions over the mobile phone conversation' when he does not want to draw attention from colleagues nearby. The civil servant adopts different strategies in dealing with different situations so as to secure his self-centred position:

> If I'm meeting with my superior, I often turn off my mobile phone or switch it to vibration. Or I quickly finish the conversation by saying that 'I am working, I will get back to you later'. The more important the

> relationship is, the more careful you have to be … I adopt language strategies to filter information. Sometimes, I only listen to her without saying too much.

While the informants who are professionals are aware of the risk involved in leaking information in public spaces and the office, the middle-aged and married informants with lower SES do not care that much. The 45-year-old waitress whose network is relatively restricted to traditional ties claims that she in fact has little to hide from others: 'I do not mind whether my friends and colleagues will listen to my conversation on the mobile phone or not in my workplace. It is not necessary for me to avoid them.' In short, the line between the public space and private space is blurred because of the diffusion of the mobile phone. However, the privatization of public space is class specific. While professional informants are more likely to privatize the public space by imagining that they are in private quarters, blue-collar workers just do it without using their imagination. In both cases, it shows liberation from social sanctions in public spaces.

Mobile phone communication in the home

An individual's home is supposed to be the most private place where one feels most comfortable and relaxed. However, our interviews reveal that informants with different socio-demographic backgrounds act differently when it comes to privacy at home. In other words, personal spaces among family members are redefined at home because of the use of the mobile phone.

The mobile phone may facilitate the creation of personal spaces at home by personalizing the family space. The young and single informants seem particularly versatile at this. The strategies they adopt include using a mobile phone rather than the fixed phone so that the family space can be easily manipulated, and communicating with intimate ties through SMS when other family members are present. The 28-year-old technician who lives with his parents tells us:

> My parents are not able to hear my mobile phone conversations because I shut the door of my room and hide under my quilt. … I often keep my mobile phone conversations away from them. I walk to another room or the balcony when speaking on the mobile phone, you know.

In respect of redefining personal space by using SMS, the younger and unmarried informants claim that they prefer communicating with the opposite sex through SMS because it is personal and silent. The 26-year-old female clerk communicates with her boyfriend by SMS when she is at home: 'I live with my parents, so it's inconvenient for me to chat on the fixed phone. I communicate with my boyfriend by SMS at night. … SMS is quiet and silent, so nobody knows whom you are communicating with.'

The younger and unmarried mobile phone users are able to create a personal space in which communication with personal intimate ties becomes possible at home. It might be necessary to point out that a younger and unmarried person could create their own privacy by using the fixed phone in their room when needed. However, the necessary privacy is only ensured if the room phone is a separate line. This is rarely the case for a Chinese family. So fixed phone communication in a Chinese family may be private but not personal when family members are present. The mobile phone allows personal conversations because of its capacity to let its users manipulate or personalize the family space.

The mobile phone and personal space are even more significant to married informants who are white-collar workers or professionals for they are more likely to have extramarital affairs. Indeed, the mobile phone is able to help personalize the family space. For instance, the male associate professor tells us that he communicates with his mistress through short messages when his wife is with him at home:

> Most of my mobile phone conversations, short messages, and MSN are with my mistress. It has become a routine, an indispensable part of my life. She has become a member of my family in my thoughts. I will surely feel low if I receive no messages from her one whole day. At least I want to know how things are going with her every day.

The 26-year-old married female who works part time at home employs both phone and SMS in order to maintain her intimate relationship with her lover who used to be her teacher:

> He never called me at the weekend when my husband was home. We talked freely only when my husband left for work. … I communicate with him very often by SMS. It gives me a sense of warmth and romance. I feel that we have gone back to my old school days when communicating with him by SMS.

Most of our informants who are married or have steady boyfriends or girlfriends claim that they do not check each other's short messages. The 37-year-old female manager says: 'Our mobile phones are openly placed at a desk or somewhere. We do not check each other's mobile phones. At least I do not check his and he doesn't either.' And a newly married female informant, aged 22, tells us her experience like this:

> I seldom check my hubby's mobile phone. But sometimes I may want to collect the funny short messages [in his mobile phone], and I will surely ask for his permission before I see them. I won't do it unless he allows it.

Nevertheless, taking precautionary measures seems necessary to some of the informants. For instance, the male associate professor always deletes all the short messages from his mistress the moment he receives and reads them in spite of the fact that his wife and son never check his short messages. These messages are 'just like a land mine that might lead to the breakdown of my family'. Our interviews show that deleting messages is a relatively common phenomenon among informants who are professionals or white-collar workers regardless of their marital status. This is understandable, for some of the spouses and fiancées have secretly checked their partner's messages for reassurance.

As far as the married informants on lower incomes in Beijing are concerned, personal space is not affected by mobile phone communication, for they do not have extramarital ties. To them, family is still family. They still regard the home as the most relaxing place. The bicycle repairman tells us that his wife gives him a sense of trust and warmth: 'I feel that I can talk about anything at home. … My wife doesn't mind what I say on the phone.'

While home is still home for the blue-collar workers, personal space is redefined and enlarged at home for the professionals or white-collar workers due to the powerful connectivity of the mobile phone.

In sum, while the informants privatize the public spaces by imagining that they are in private quarters when on the phone, they personalize the private spaces such as the family apartment by redefining 'personal'. However, the degree of privacy needed depends upon our informants' socio-demographic characteristics. Indeed, the mobile phone enables our informants to find the necessary privacy or communicate privately or personally because it is not only mobile but also voice- and text-based. Perhaps we may say that the mobile phone means privacy by its very nature. We need to communicate privately or personally because we are afraid of social sanctions sometimes. The mobile phone frees us from such sanctions to some degree. Thus, we may have a better understanding of the changes in social relations discussed in the previous section. Indeed, the mobile phone very much favours 'person-to-person' connectivity.

Discussion and concluding remarks

Spatial changes affect social structures to a large degree, for spaces are closely related to social relations. Among Western studies on social transitions and new media, the concept of 'networked individualism' seems particularly useful (Castells 2000; Haythornthwaite and Wellman 2002; Wellman 2001, 2002). As demonstrated above, this chapter attempted to explore the specific role of the mobile phone in changing social structures in today's Beijing. We have shown that the mobile phone further frees its users from space constraints as far as social networking is concerned. That is, it allows the users to privatize public space by creating a sense of imagined privacy in front of present others and to personalize private space by redefining the boundaries at home in order to maintain personal and extraneous ties (cf. Fortunati 2002a).

With the adoption and diffusion of the mobile phone, the public spaces are privatized and the private spaces are personalized, which means the development of 'networked individualism', or a transition to 'personal community', as reviewed above. As our data have suggested, the adoption of the mobile phone promotes the emergence of a personal community where an individual acts as the centre of their network. However, this is not to say that our informants' traditional networks have disappeared. With the help of the mobile phone, many informants, especially the younger or the middle-aged with higher incomes, have developed skills in dealing with different social relations belonging to either the traditional network or the individualism-oriented community. The large collection of phone numbers stored in the mobile devices of most informants except those on lower incomes suggests to a certain extent the social transition from the 'spatially compact and densely knit' communities (see Barthélemy and Contamine 1985; Thébert 1985; Ward 1999) to the more 'sparsely knit' communities transcending physical confines (Haythornthwaite and Wellman 2002).

It has been argued that the Internet lures people away from their family and community (Barlow 1995; Hiltz and Turoff 1993; Kling 1996; Kraut *et al.* 1998; Nie 2001; Nie and Erbring 2000; Rheingold 1994). But it seems that the mobile phone helps maintain kinship ties. Our informants' mobile phone communications with their relatives appear to be frequent and affective in nature (cf. Law 2005). This is particularly true for parent–child relationships. But, our data also suggest that mobile communication between husbands and wives is frequent yet instrumental to some degree. The mobile phone, and particularly SMS, has enabled the construction of personal space between husband and wife, at least to a certain extent. We have emphasized and elaborated on the concept of 'personalizing private space' in this chapter. Private space such as the family is redefined with the help of the mobile phone. It is claimed by almost all the married informants that they do not check their spouse's short messages and that they regard the messages as personal and private. Co-presence no longer ensures loyalty or fidelity for couples even at home.

Almost all our married informants of higher SES have maintained at least one extra-marital romantic relationship with the assistance of the mobile phone. The point is that marriages need private space to grow. Mobile phone communication has changed the ecology by enabling the personalization of the private space fenced in by walls and gates. While the mobile phone can be used for micro-coordination within families, it is also likely to be a catalyst for the weakening of ties between spouses as far as our informants are concerned. Indeed, the mobile phone has provided seemingly unlimited power in person-to-person connectivity.

In urban China, best friends are usually classmates, workmates, comrades-in-arms, or neighbours. They often share each other's life history and other life dimensions. Nevertheless, for almost all the informants except the younger and unmarried female users and the married female users on lower incomes, having 'best friends' does not necessarily mean frequent face-to-face communication or mediated communication, such as via the mobile phone, regardless of distance. This is similar to what Wellman (1979: 1214) reported on East Yorker long-distance intimate ties decades ago: 'Their infrequent contact ratifies the tie, and a potential is retained for more intensive use when needed.' Compared with our findings, the difference is also obvious. Our informants on higher incomes claim that their infrequent contact with their best friends is due to the fact that they are busy, rather than because of problems of distance. In fact, best friends are local for most of our younger male and middle-aged informants on higher incomes. They occasionally affirm the ties when necessary. The multi-dimensional relationship between or among best friends tends to shrink to needs-based friendship. In this sense, we suggest that the ties between or among best friends are also weakened for the younger male and middle-aged informants on higher incomes in an age of mobile phone communication.

The mobile phone functions differently for different social ties. We have discussed the dual role of the mobile phone in maintaining family ties. We have also suggested that the mobile phone does not favour the ties between or among best friends. However, this is only part of the communication pattern that has emerged as far as the mobile phone is concerned. Indeed, most of our informants, especially the younger ones or those on higher incomes, have more and intensive mobile phone communications with others besides kinship and love ties.

The mobile phone is not necessarily used to initiate new ties. The Internet is employed for that because the Internet creates more opportunities for communication with others (Constant *et al.* 1996; Dobashi 2005; Feldman 1987). Our findings show both media are used for developing new friendships. The difference is that the Internet is used to join a particular 'circle' where different people come to know each other, while mobile communication is more likely to be used to maintain the tie with a specific friend or the person selected.

As Fortunati puts it, 'the purpose of the mobile is to be reachable … by those with whom we want to communicate – intimate friends or selected others …' (Fortunati 2002a: 51). Matsuda (2005) has also revealed this 'selective sociality' among Japanese mobile users. Our research found that the younger informants are more likely to develop friendships with the newly met and keep regular contact with them for leisure-time entertainment such as singing in Karaoke bars or group holidays. Some younger and unmarried informants claim that they preserve regular communication with their friends living in another city or country mainly by SMS and the Internet for intimate and emotional exchanges.

Our informants use the mobile phone to communicate with their kinship ties, romantic friends, and other social ties intensively. Such communication occurs not only in private space such as the home but also in public space. Previous studies suggest that the use of the mobile phone in public contributes to the blurring of the boundary between the public and the private (Castells *et al.* 2004; Kim 2002). While some of our informants are more concerned about talking on the mobile phone in public places, they are all privatizing the public space by either imagining that they have the necessary privacy or by simply ignoring the surroundings when they take part in mobile phone communication.

While the fixed phone enables strong ties to go beyond the neighbourhood, the mobile phone seems to liberate users from physical space in all dimensions. By means of the mobile phone, we can turn public space into private space or private space into personal space. When we are less and less physically and mentally constrained in respect to social relations and sanctions, our social behaviour features networked individualism. Speaking of today's family transition, Giddens (2002) argues that 'Are you in a relationship?' is a more important question than 'Are you married?' It is probable that 'Are you with somebody right now?' is the most important question in an age of mobile phone communication. Indeed, the medium enables the construction of personal space as well as individual actions.

Note

1 The author would like to thank Ms Bo Gai and Ms Li Li for contributing to the very first draft of this chapter three years ago when the two of them were PhD students in the School of Journalism and Communication, Peking University.

References

Agre, P. (2001) 'Changing places: contests of awareness in computing', *Human–Computer Interaction*, 16: 177–92.

Barlow, J. (1995) 'Is there a there in cyberspace?', *Utne Reader*, March–April: 50–56.

Barthélemy, D. and Contamine, P. (1985) 'The use of private space', in G. Duby (ed.) *A History of Private Life*, Cambridge, MA: Belknap Press.

Beijing Municipal Bureau of Statistics (2009) *Beijing Statistics Yearbook, 2008*. Released January 2009. Online. Available: www.bjstats.gov.cn/tjnj/2008-tjnj/ (accessed 24 February 2009).

Boneva, B. and Kraut, R. (2002) 'Email, gender, and personal relationships', in B. Wellman and C. Haythornthwaite (eds) *The Internet in Everyday Life*, Oxford: Blackwell.

Castells, M. (2000) *The Rise of the Network Society*, 2nd edn, Oxford: Blackwell.

Castells, M., Fernandez-Ardeveol, M., Qiu, J. and Sey, A. (2004) 'The mobile communication society: a cross-cultural analysis of available evidence on the social uses of wireless communication technology', paper presented at the International Workshop on Wireless Communication Policies and Prospects: A Global Perspective, Los Angeles, October.

Chen, W., Boase, J. and Wellman, B. (2002) 'The global villagers: comparing Internet users and uses around the world', in B. Wellman and C. Haythornthwaite (eds) *The Internet in Everyday Life*, Oxford: Blackwell.

Coffey, S. and Stipp, H. (1997) 'The interactions between computer and television use', *Journal of Adolescent Research*, 37: 61–67.

Constant, D., Kiesler, S. and Sproull, L. (1996) 'The kindness of strangers: the usefulness of electronic weak ties for technical advice', *Organization Science*, 7: 119–35.

Dobashi, S. (2005) 'The gendered use of *Keitai* in domestic contexts', in M. Ito, D. Okabe and M. Matsuda (eds) *Personal, Portable, Pedestrian: mobile phones in Japanese life*, Cambridge, MA: MIT Press.

Fei, X. (2006) *Xiangtu Zhongguo* [From the Soil: foundations of Chinese society], 2nd edn, Beijing: Beijing Press.

Feldman, M. (1987) 'Electronic mail and weak ties in organizations', *Office: Technology and People*, 3: 83–101.

Fortunati, L. (2000) 'The mobile phone: new social categories and relations', paper presented at the seminar *Sosiale Konsekvenser av Mobiltelefoni*, Oslo, June.

—— (2002a) 'Italy: stereotypes, true and false', in J. Katz and M. Aakhus (eds) *Perpetual Contact: mobile communication, private talk, public performance*, Cambridge: Cambridge University Press.

—— (2002b) 'The mobile phone: towards new categories and social relations', *Information, Communication, and Society*, 5: 514–28.

Gergen, K. (2002) 'The challenge of absent presence', in J. Katz and M. Aakhus (eds) *Perpetual Contact: mobile communication, private talk, public performance*, Cambridge: Cambridge University Press.

Giddens, A. (1990) *The Consequences of Modernity*, Stanford, CA: Stanford University Press.

—— (2002) *Runaway World: how globalisation is reshaping our lives*, London: Profile Books.

Habuchi, I. (2005) 'Accelerating reflexivity', in M. Ito, D. Okabe and M. Matsuda (eds) *Personal, Portable, Pedestrian: mobile phones in Japanese life*, Cambridge, MA: MIT Press.

Hampton, K. and Wellman, B. (2002) 'The not so global village of Netville', in B. Wellman and C. Haythornthwaite (eds) *The Internet in Everyday Life*, Oxford: Blackwell.

Hashimoto, K., Hashimoto, Y., Ishii, K., Nakamura, I., Korenaga, R., Tsuji, D. and Mori, Y. (2000) 'Survey research on uses of cellular phones and other communication media in 1999', *The Research Bulletin of the Institute of Socio-Information and Communication Studies*, 14: 83–192 (in Japanese).

Haythornthwaite, C. (2001) 'Strong, weak, and latent ties and the impact of new media', *The Information Society*, 18: 385–401.

Haythornthwaite, C. and Wellman, B. (2002) 'The Internet in everyday life: an introduction', in B. Wellman and C. Haythornthwaite (eds) *The Internet in Everyday Life*, Oxford: Blackwell.

Hiltz, S. and Turoff, M. (1993) *The Network Nation: human communication via computer*, 2nd edn, Cambridge, MA: MIT Press.

Howard, P., Rainie, L. and Jones, S. (2002) 'Days and nights on the Internet', in B. Wellman and C. Haythornthwaite (eds) *The Internet in Everyday Life*, Oxford: Blackwell.

Ishii, K. (2004) 'Internet use via mobile phone in Japan', *Telecommunications Policy*, 28: 43–58.

Ito, M. and Okabe, D. (2005) 'Technosocial situations: emergent structuring of mobile e-mail use', in M. Ito, D. Okabe and M. Matsuda (eds) *Personal, Portable, Pedestrian: mobile phones in Japanese life*, Cambridge, MA: MIT Press.

Katz, J. and Aspden, P. (1997) 'A nation of strangers?', *Communications of the ACM*, 40: 81–86.

Katz, J. and Rice, R. (2002) 'Syntopia: access, civic involvement and social interaction on the net', in B. Wellman and C. Haythornthwaite (eds) *The Internet in Everyday Life*, Oxford: Blackwell.

Kim, S. (2002) 'Korea: personal meanings', in J. Katz and M. Aakhus (eds) *Perpetual Contact: mobile communication, private talk, public performance*, Cambridge: Cambridge University Press.

Kling, R. (1996) 'Analysis of electronic support groups for recovering addicts', *Interpersonal Computing and Technology*, 2: 47–56.

Kraut, R., Patterson, M., Lundmark, V., Kiesler, S., Mukhopadhyay, T. and Scherlis, W. (1998) 'Internet paradox: a social technology that reduces social involvement and psychological well-being?', *American Psychologist*, 53: 1017–31.

Laurier, E. (2001) 'Why people say where they are during mobile phone calls', *Environment and Planning D: Society and Space*, 19(4): 485–504.

Law, P.L. (2005) 'Mobile communication, mobile networks, and mobility: the case of migrant workers in southern China', paper presented at the International Conference on Mobile Communication and Asian Modernities II: Information, Communications Tools and Social Changes in Asia, Beijing, October.

Ling, R. (2000) 'Direct and mediated interaction in the maintenance of social relationships', in A. Sloane and F. van Rijn (eds) *Home Informatics and Telematics: information, technology and society*, Boston, MA: Kluwer.

Matei, S. and Ball-Rokeach, S. (2002) 'Belonging in geographic, ethnic and Internet spaces', in B. Wellman and C. Haythornthwaite (eds) *The Internet in Everyday Life*, Oxford: Blackwell.

Matsuda, M. (2005) 'Mobile communication and selective sociality', in M. Ito, D. Okabe and M. Matsuda (eds) *Personal, Portable, Pedestrian: mobile phones in Japanese life*, Cambridge, MA: MIT Press.

MIIT [Ministry of Industry and Information Technology] (2011) 'China telecommunication statistics 2010', released 26 January 2011. Online. Available: www.miit.gov.cn/n11293472/n11293832/n11294132/n12858447/13578942.html (accessed 17 February 2011).

Murtagh, G. (2002) 'Seeing the "rules": preliminary observations of action, interaction and mobile phone use', in B. Brown, R. Harper and N. Green (eds) *Wireless World: social and international aspects of the mobile age*, New York: Springer.

Nie, N. (2001) 'Sociability, interpersonal relations, and the Internet: reconciling conflicting findings', *American Behavioral Scientist*, 45: 420–35.

Nie, N. and Erbring, L. (2000) *Internet and Society: a preliminary report*, Stanford, CA: Stanford Institute for the Quantitative Study of Society, Stanford University.

Palen, L., Salzman, M. and Youngs, E. (2001) 'Discovery and integration of mobile communications in everyday life', *Personal and Ubiquitous Computing*, 5: 109–22.

Quan-Haase, A., Wellman, B., Witte, J. and Hampton, K. (2002) 'Capitalizing on the net: social contact, civic engagement, and sense of community', in B. Wellman and C. Haythornthwaite (eds) *The Internet in Everyday Life*, Oxford: Blackwell.

Rheingold, H. (1994) *The Virtual Community: homesteading on the electronic frontier* (revised edition), London: MIT Press.

Robinson, J. and Godbey, G. (1997) *Time for Life*, University Park: Pennsylvania State University Press.

Robinson, J., Kestnbaum, M., Neustadtl, A. and Alvarez, A. (2002) 'The Internet and other uses of time', in B. Wellman and C. Haythornthwaite (eds) *The Internet in Everyday Life*, Oxford: Blackwell.

Rong, W. (2001) 'Shehui Xingbie Yu Shenmei' [Gender and the aesthetic taste], in Q. Li (ed.) *Ban Ge Shiji De Funv Fazhan: Zhongguo Funv 50 Nian Lilun Yantaohui Lunwenji* [Half Century of Women's Development: a collection of papers from theory conferences on Chinese women for the past 50 years], Beijing: Dangdai Zhongguo Chubanshe [Contemporary China Press].

Su, H. (2004) *Duochong Shijiao Xiade Shehui Xingbeiguan* [Perspectives on Gender], Shanghai: Shanghai Daixue Chubanshe [Shanghai University Press].

Thébert, Y. (1985) 'Private life and domestic architecture in Roman Africa', in P. Veyne (ed.) *A History of Private Life*, Cambridge, MA: Belknap Press.

Tsuji, D. and Mikami, S. (2001) 'A preliminary student survey on the e-mail uses by mobile phones', paper presented at JSICR (Japanese Society of Information and Communication Research) annual conference, Tokyo, June.

Ward, P. (1999) *A History of Domestic Space: privacy and the Canadian home*, Vancouver: UBC Press.

Wellman, B. (1979) 'The community question: the intimate networks of East Yorkers', *The American Journal of Sociology*, 84: 1201–31.

—— (1988) 'Structural analysis: from method and metaphor to theory and substance', in B. Wellman and S. Berkowitz (eds) *Social Structures: a network approach*, Cambridge: Cambridge University Press.

—— (1999) 'The network community', in B. Wellman (ed.) *Networks in the Global Village*, Boulder, CO: Westview.

—— (2001) 'Physical place and cyber place: the rise of personalized networking', *The International Journal of Urban and Regional Research*, 5: 227–52.
—— (2002) 'Little boxes, glocalization, and networked individualism', in M. Tanabe, P. Besselaar and T. Ishida (eds) *Digital Cities II: computational and sociological approaches*, Berlin: Springer-Verlag.
Wellman, B. and Gulia, M. (1999) 'Net surfers don't ride alone: virtual communities as communities', in B. Wellman (ed.) *Networks in the Global Village*, Boulder, CO: Westview Press.
Wellman, B. and Leighton, B. (1979) 'Networks, neighborhoods and communities', *Urban Affairs Quarterly*, 14: 360–90.
Wellman, B. and Tindall, D. (1993) 'Reach out and touch some bodies: how telephone networks connect social networks', *Progress in Communication Science*, 12: 63–94.
Wellman, B. and Wortley, S. (1990) 'Different strokes from different folks: community ties and social support', *American Journal of Sociology*, 96: 558–88.
Ye, W. and Lin, Q. (1998) 'Dangdai Zhongguo Lihun Taishi He Yuanyin Fenxi' [An analysis on the situation and causes of divorce in contemporary China], *Renkou Yu Jingji* [Population and Economics], 3: 22–28.
Yuen, S.P., Law, P.L. and Ho, Y.Y. (2004) *Marriage, Gender, and Sex in a Contemporary Chinese Village*, New York: M.E. Sharpe.

Index

Note: Page numbers for figures and tables appear in **bold**.